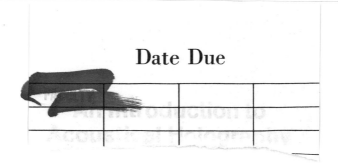

An Introduction to Acoustical Holography

B. P. Hildebrand

Battelle Memorial Institute
Pacific Northwest Laboratories
Richland, Washington

and

B. B. Brenden

Holosonics, Inc.
Richland, Washington

A PLENUM/ROSETTA EDITION

Library of Congress Cataloging in Publication Data

Hildebrand, B P   1930-
    An introduction to acoustical holography.

    "A Plenum/Rosetta edition."
    Includes bibliographical references.
    1. Acoustic holography. I. Brenden, B. B., joint author. II. Title.
    [QC244.5.H55   1974]   535'.4   73-23037
    ISBN 0-306-20005-8 (pbk.)

A Plenum/Rosetta Edition
Published by Plenum Publishing Corporation
227 West 17th Street, New York, N.Y. 10011

First paperback printing 1974

United Kingdom edition published by Plenum Press, London
A Division of Plenum Publishing Company, Ltd.
4a Lower John Street, London W1R 3PD, England

This book is dedicated with great affection to our wives,

Thelma Hildebrand and Lavelle Brenden,

for their unquestioning loyalty and support.

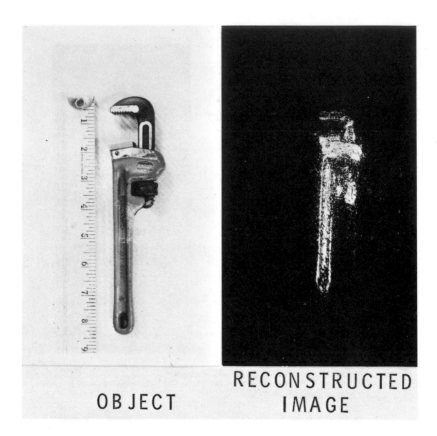

OBJECT   RECONSTRUCTED IMAGE

Pipe wrench—This acoustical image of the pipe wrench was produced from a source-receiver scanned hologram having an aperture of 15 x 15 cm and a line density of 33 lines/cm. The pipe wrench was insonified with 5.1 MHz sound produced by a 1-in.-diameter focused transducer.

Preface

Since the first papers by E. N. Leith and J. Upatnieks on the subject of holography appeared in 1961, there has been a virtual explosion of research activity in the field. More than 500 papers and articles on holography have appeared in the last ten years. Many applications of holography have been proposed, and some of these are beginning to enter the realm of usefulness. One of the applications that appears to hold great promise is acoustic imaging by means of holography. The first papers on this subject appeared in 1966, but already research activity in the field is burgeoning. Three symposia wholly devoted to acoustical holography have been held and the papers published in book form.

The purpose of this book is to bring together the results of research in acoustical holography, some of it as yet unpublished, under one cover so that workers in holography, nondestructive testing, medical imaging, underwater imaging, and seismic exploration can decide whether this new technique can be useful to them.

The treatment of the book requires some knowledge of differential equations and diffraction theory, but is kept as simple as possible. The first chapter includes an historical sketch of the development of holography since its invention in 1948 by Gabor. The second chapter is devoted to the development of the holographic imaging equations using the approach first used by Meier. The third chapter serves as an introduction to acoustics for those readers unfamiliar with these concepts. The following three chapters describe the various methods for obtaining and reconstructing acoustical holograms with particular emphasis on the two most-developed methods; liquid-surface and scanning. Chapter seven provides brief descriptions of other techniques that have appeared in the literature with experimental results, where available. The last chapter describes possible applications of acoustical holography to ocean surveillance, medicine, nondestructive testing, seismic exploration, and nuclear technology. Wherever possible, experimental results are shown.

We realize that a book published while research is still going on at a frantic pace is often obsolete by the time it comes out in print. Therefore, we have endeavored to use experimental results that were obtained concurrently with the writing in order that at publication it be as up-to-date as possible. All such results, even for chapters discussing holography in general, were obtained with acoustical radiation at a frequency of 3 MHz unless otherwise noted.

Much of the research reported in this book was performed at the Pacific Northwest Laboratories of the Battelle Memorial Institute, under the sponsorship of the Holotron Corporation of Wilmington, Delaware. We thank our many colleagues for allowing us to draw upon their work, in particular R. B. Smith, D. R. Hoegger, T. J. Bander, V. I. Neeley, S. C. Keeton, F. V. Richard, G. Langlois, H. Toffer, K. A. Haines, and D. S. St. John. Special thanks is due H. D. Collins and R. P. Gribble who provided a majority of the experimental results. In addition, one of us (B. P. H.) wishes to thank E. N. Leith for introducing him to this fascinating field while a student at the University of Michigan.

Thanks is due to Battelle Memorial Institute for providing the creative atmosphere and support of the Battelle Seattle Research Center during the writing of this monograph. Battelle-Northwest has also supported this effort significantly. We wish to thank G. J. Dau for his personal interest in this work and Mrs. Janice Sletager for editing the manuscript and coordinating the graphics and reproduction requirements.

Finally, we acknowledge the largesse of the University of British Columbia in the person of Miss Kathy Hardwick who typed the final manuscript and Holosonics Inc., for supplying some of the photographs for the liquid-surface holography results.

B. P. HILDEBRAND
B. B. BRENDEN

December 1971

Contents

Chapter 1

Introduction

1.1. FUNDAMENTAL CONCEPTS

Holography is a synthesis, for the purpose of recording and displaying an image, of two venerable branches of optics. The two branches referred to are interferometry, used in recording the hologram, and diffraction, used to display the image.

Although interference and diffraction are described in any textbook dealing with optics, they are always treated as separate topics. Only recently were the two combined to form a single branch of optics that includes holography and optical data processing. The impetus for this synthesis came from engineers and physicists working in the area of communication. In communications the analogies to interference and diffraction are modulation and demodulation. From this point of view it becomes obvious that information stored in an interference pattern (modulation) can be recovered as a result of diffraction of light by the recorded interference pattern (demodulation).

We now undertake a simplified discussion of holography based upon interference and diffraction and assume that the reader has sufficient background to accept the existence of these phenomena. Interference, in this context, is best discussed with the aid of an interferometer of the Michelson variety as sketched in Fig. 1.1. A plane monochromatic wave is split into two equal amplitude plane waves which then traverse separate paths until they are recombined. If the mirrors and beamsplitters are perfect, and if the optical path lengths are equal, the output beam will be exactly like the input beam except for the loss in amplitude incurred at the beamsplitter. A screen placed in the output beam will therefore show uniform intensity.

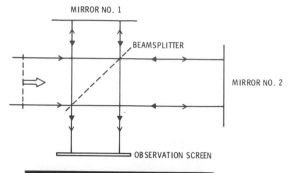

Fig. 1.1. The photograph represents the interference pattern obtained from the acoustical equivalent of the Michelson interferometer shown above it. The beam-splitter is replaced by a phase detector in which the signal from a scanning receiver is mixed with an electronic reference, and the observation screen by a film illuminated by a scanning light source modulated by the output of the phase detector.

If one of the mirrors is not quite parallel to the wave front, due to a misalignment, the reflected wave from it will recombine with the other beam to form a linear system of fringes as shown in Fig. 1.2. Destructive interference occurs wherever the path length difference between the wave fronts is an odd multiple of half the wavelength.

If one of the mirrors is replaced with a concave parabolic reflector, the interference pattern we would expect to see is shown in Fig. 1.3. Note that the ring pattern shows decreasing spacing with increasing distance from the center. The distance between rings as a function of ring number k (as counted from the center of the pattern) is approximately

$$\sqrt{R\lambda}\ (\sqrt{2k+3} - \sqrt{2k+1})$$

where R is the radius of the spherical wave and λ is the wavelength of the radiation. This type of pattern is called the Fresnel zone pattern. If we impart a tilt to the spherical wave we obtain a partial Fresnel zone pattern as shown in Fig. 1.4.

When an aperture of some kind is interposed in a collimated beam of light we might expect to see, on the screen, a pattern of light sharply defined by the shadow of the aperture. Looking closely, however, the observer notes that some light exists in the shadow zone. The study of this phenomenon is known as diffraction theory. The essence of the theory, known as Huygens' principle, is that each point on a wave front can be considered as an elementary point source radiating a spherical wave front. When the

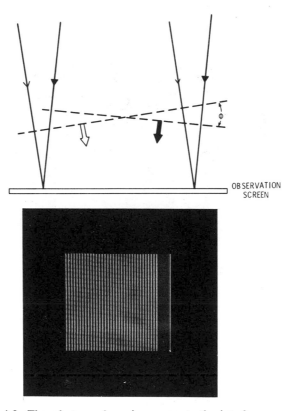

OBSERVATION
SCREEN

Fig. 1.2. The photograph again represents the interference pattern obtained from the acoustical equivalent of a Michelson interferometer. The difference between this result and that shown in Fig. 1.1 is a linear-spatial phase shift of the reference signal resulting in linear interference fringes.

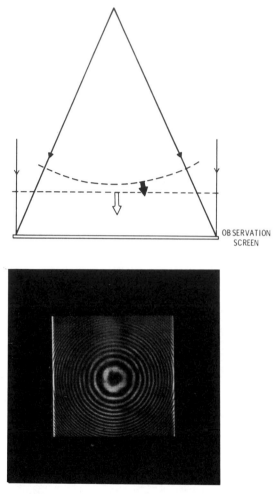

OBSERVATION
SCREEN

Fig. 1.3. The Fresnel zone pattern was generated by the interference of a spherical wave with a simulated plane wave in the manner described in Fig. 1.1.

elementary wavelets are summed in the prescribed manner, the result is identical to the actual wave front farther downstream. This principle is useful in analyzing the behavior of light in the presence of various types of apertures or obstructions. We now proceed to do this using as apertures the interference patterns we considered in the preceding paragraphs.

For simplicity we will assume that the interference patterns are binary, i.e., transparent or opaque. In actuality the transmission varies proportionally to $\sin^2$. First, we consider the pattern generated in Fig. 1.2. A piece of film containing this pattern is interposed in a collimated monochromatic light beam as shown in Fig. 1.5. In this figure we consider a single wavelet in the center of each transparent space. Note that each circle represents an equiphase surface advanced in phase by 2π from its neighbor to the left.

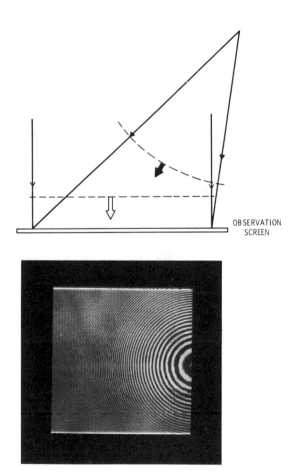

Fig. 1.4. The Fresnel zone pattern has been displaced by moving the spherical source transducer out of the aperture.

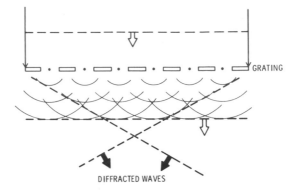

Fig. 1.5. The upper figure represents Huygens' construction from a plane grating illustrating the diffracted
plane waves. The lower picture is a photograph at the
focal plane of a lens showing the focused diffracted
waves from the interference pattern generated in Fig. 1.2.
Due to the square-wave nature of the grating second-
order diffraction is also present.

The spatial separation of each such surface from the next is one wavelength λ. We can draw a plane phase front tangent to the elementary wavelet phase fronts, three of which are shown in Fig. 1.5. For the binary pattern
we can draw many more plane wave fronts at ever increasing angles. If we
had the actual interference pattern (i.e., $\sin^2$ transmission variation), all but
these three would be attenuated. Comparing Fig. 1.5 with Fig. 1.2, we see

that we have reconstructed the beams that made the interference pattern in the first place, with one exception; we have an extra beam at the conjugate angle. More will be said about this later.

The second example of an interference pattern was the Fresnel zone pattern (Fig. 1.3). Figure 1.6 shows what happens when this is illuminated. The Fresnel zone pattern is, in effect, a locally linear grating acting much the same as our first example except that the spacing varies as described previously. Hence, the locally diffracted light changes in direction and the total effect provides a spherical wave front, again duplicating the beams that caused the original interference pattern. Again, we have an extra wave of opposite curvature.

The last example we considered was that of an offset Fresnel zone pattern (Fig. 1.4). For completeness we show in Fig. 1.7 the result of introducing this pattern in a beam of light, although by now the reader no doubt anticipates the consequences.

The significance of the foregoing discussion is that interferometry has always had the capability of recording a wave front and later reconstructing it. It is remarkable that this fact was not grasped much earlier than it was. Our three examples of interference patterns were carefully chosen to lead the reader to understand, on physical grounds, how and why holography works. The first example (Fig. 1.2) was intended to demonstrate both interference and diffraction. The second example (Fig. 1.3), a slightly more complicated situation, was chosen to show how Gabor holography works and why it is not too successful. Using the Gabor approach, one is reconstructing three beams of light, all occupying the same volume. That is, if we consider Fig. 1.6 and imagine that we are looking for a replica of the spherical wave used in making the interferogram, we are also forced to look at the plane wave and the converging extraneous wave, both of which tend to mask the light from the desired wave.

The third example, Fig. 1.4, shows how this difficulty is overcome by using an offset in the spherical beam. The same result can be produced by bringing the plane reference beam in at an angle, although in this case the interference pattern is not a circular Fresnel pattern. In this case, all three diffracted beams are separated in space, thus completely solving the problem of overlap.

To make these discussions conform to what one generally likes to think of as holography, one further step is required; namely, to record and display an image of a complicated object. This step only requires that we accept Huygens' principle which states that any arbitrary wave front representing the light reflected from a complicated object can be considered to

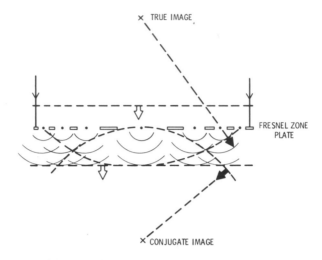

Fig. 1.6. The upper figure shows the Huygens' construction for a centered Fresnel zone plate. The photograph is again taken at the focal plane of a lens. Therefore the bright central spot represents the focused plane wave and the rectangles of light the two out of focus spherical waves. Note that all are superimposed.

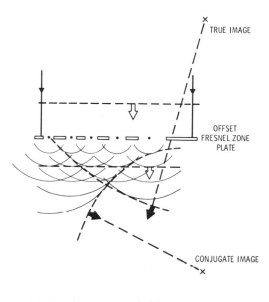

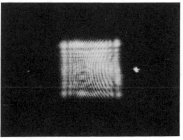

Fig. 1.7. The upper figure shows the Huygens' construction for an off-center Fresnel zone plate. The
photograph clearly illustrates the separation of the
three waves. The bright point is the conjugate image
(the focal plane of a lens) the rectangle is the unfo-
cused plane wave. Note the physical separation. The
true image is not visible since it is too much out of
focus and hence underexposed.

be the sum of a large number of point sources distributed over that wave
front. Since we have shown that we can image a single point or spherical
wave, it is reasonable to suppose that a distribution of points can also be
imaged.

The foregoing theory depends upon one overriding principle; that is
that the two beams must be capable of interfering. This, in turn requires
a property of light, known as coherence, about which much has been written,

and we will not belabor here. In this book we will *always* assume that the radiation is coherent.

Since this book is about acoustical holography we must relate this to the optical variety described above. It should be noted that interference and diffraction are phenomena common to all forms of energy whose propagation can be described in terms of wave motion. All that is required is the ability to record an interference pattern in the particular type of radiation we happen to be using. One does not always have a convenient recording medium like film. Hence, it is necessary to find a means of recording the interference pattern in the particular radiation one is working with, and then to use this pattern to diffract light into a replica of the original wave. In this way it is possible to produce a visual image of an object irradiated with invisible energy. Acoustical energy is one type of radiation that has extremely useful properties, the most important being its ability to penetrate solids and liquids. Thus, one can image objects in opaque fluids, identify flaws in metals, or perhaps detect cancerous growth in tissues.

The purpose of this book, then, is to summarize the progress to date, of research in the use of holography for acoustical imaging and to try to suggest the areas of application in which it will be most useful.

1.2. HISTORICAL DEVELOPMENT

Having sketched the physical principles upon which holography is based, it seems appropriate to briefly trace the history of its development. As we mentioned in Section 1.1, holography is a synthesis, only recently made, of interference theory and diffraction theory. Of these two subjects of classical optics, diffraction is the oldest. It was first discussed by Grimaldi [1] in 1665 and was later put on a sound mathematical foundation by Huygens [2] in 1678. At about this time, Newton concluded that light was propagated as particles (mechanical corpuscular theory) and his great authority effectively stifled further research into the wave theory of light.

It remained for Young [3] to demonstrate interference and propound the wave theory explanation. However, it was not until Fresnel [4] in 1816 performed a detailed analysis of Huygens' principle and Young's interference principle by means of wave theory, that the mechanical corpuscular theory was abandoned.

Interferometry rapidly became a major tool for precision measurements. Michelson [5], for instance, used it to disprove the ether drift theory. It was also used extensively to test for imperfections in lenses and precision mirrors. It is this application that comes closest to holography.

In the late 1940's Gabor [6], who was working on the improvement of the electron microscope, proposed a two-step method of imagery. At that time the electron microscope suffered from severe spherical aberration. Gabor's idea was to dispense with the lens by recording the electron diffraction pattern of the object and use this pattern to reconstruct an image, in this way circumventing the aberration. It is interesting to note that this particular application of holography has never come to pass. Rather, the rapid improvement in electron beam technology overcame the initial problem of aberration.

As we described in Section 1.1, Gabor's method had some shortcomings due to the generation of an extraneous image which could not be separated from the desired one. Many attempted to correct this defect with varying degrees of success [7]. It was not until 1962 that Leith [8], utilizing his experience in communications theory, succeeded in separating the images by the simple expedient of interfering the two beams with an angular separation between them rather than on the same axis. This, as we saw in Section 1.1, caused the reconstructed beams to be physically separated in space.

The invention of the gas laser at the same time provided the perfect coherent source which made holography so spectacular that scientists and laymen alike were enthralled. Thus, as a result of wide publicity and the research money that became available, the technique was rapidly developed into many diverse fields.

Since Gabor's original intent was to use holography as a means of visualizing nonvisible radiation it is not surprising that researchers rapidly converged on this application. Holograms have been made with microwave and ultrasonic radiation with good success and attempts have been made using X-rays and electron beams with lesser success. The first published work on ultrasonic holography was, as far as we know, by Thurstone in 1966 [9], although it is certain that many individuals were working on it at the same time. The main problem in developing this application was that of recording, since photographic film is not normally sensitive to sound. Two main methods of investigation have been developed for recording; the point-by-point sampling method using one or more acoustical receivers [10] and the liquid–gas interface method [11]. A variety of read-out techniques have been developed and will be considered in the body of this book.

REFERENCES

1. F. M. Grimaldi, Physico-Mathesis de lumine, coloribus, et iride (Bologna, 1665).
2. Chr. Huygens, Traité de la lumière (completed in 1678, published in Leyden in 1690).
3. Th. Young, Phil. *Trans. Roy. Soc. London* xcii (1802) **12**:387; Young's Works, Vol. 1, p. 202.
4. A. J. Fresnel, *Ann. Chim. Phys.* **1**(2):239 (1816); *Œuvres*, Vol. 1, 89, 129.
5. A. A. Michelson, *Light Waves and their Uses*, University of Chicago Press (1902).
6. D. Gabor, A new microscope principle, *Nature* **161** (1948).
7. A. Lohmann, Optical single-sideband transmission applied to the Gabor microscope, *Opt. Acta* **3**:97 (1956).
8. E. N. Leith and J. Upatnieks, Reconstructed wavefronts and communication theory, *J. Opt. Soc. Am.* **52**:1123 (1962).
9. F. L. Thurstone, Ultrasound holography and visual reconstruction, *Proc. Symp. Biomed. Eng.* **1**:12 (1966).
10. K. Preston, Jr., and J. L. Kreuzer, Ultrasonic imaging using a synthetic holographic technique, *App. Phys. Lett.* **10**(5):150 (1967).
11. R. K. Mueller and N. K. Sheridon, Sound holograms and optical reconstruction, *Appl. Phys. Lett.* **9**(9):328 (1966).

Holography

2.1. RECORDING PROCESS

In Chapter 1 we established on a qualitative physical basis how a modified form of interferometry could be used to record a wave front. This record, in turn, could be used to reconstruct that wave front by the process of diffraction. This method of photography has come to be known as holography after the word "hologram" coined by Gabor to describe such an interferogram [1]. In this chapter we describe this process in mathematical terms. Although many authors use the complex notation and suppress the time dependence [2] of the radiation, we will use the real notation and retain the time-dependent terms [3]. We do this only because we wish to consider long wavelength radiation where it is not as obvious that the time dependence can be ignored as with optical radiation.

For the moment we consider two beams of radiation intersecting on a detecting plane as shown in Fig. 2.1. These two beams are derived from the same source by some kind of interferometric arrangement as discussed in Chapter 1, or in the case of acoustical waves they may come from two transducers driven by the same oscillator. We can describe the radiation at a point (x, y) on the detecting surface due to beam 1 as

$$s_1(x, y) = a_1(x, y) \cos[\omega t + \phi_1(x, y)] \qquad (2.1)$$

and the radiation due to beam 2 as

$$s_2(x, y) = a_2(x, y) \cos[\omega t + \phi_2(x, y)] \qquad (2.2)$$

where $a(x, y)$ is the amplitude of the wave, $\phi(x, y)$ is the phase of the wave,

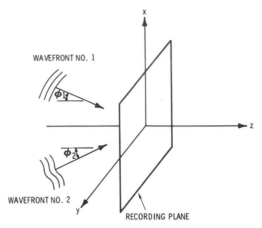

Fig. 2.1. This figure represents the way in which a hologram is made. Either one of the wavefronts can represent the object wave while the other represents the reference wave.

and ω is the temporal radian frequency of the radiation. The total amplitude at (x, y) is the sum of these two individual beams. The result is

$$s(x, y) = s_1(x, y) + s_2(x, y)$$
$$= a_1(x, y) \cos[\omega t + \phi_1(x, y)] + a_2(x, y) \cos[\omega t + \phi_2(x, y)] \quad (2.3)$$

In optical radiation, the radian frequency is so great that no detector exists which can detect the oscillation. The best that can be done is to measure intensity. The signal that is actually recorded is therefore some function of intensity, which can be written

$$I(x, y) = \langle [s(x, y)]^2 \rangle_t \quad (2.4)$$

where $\langle \ \rangle_t$ denotes a time average over many cycles of the radiation. Substituting Eq. (2.3) into Eq. (2.4) yields

$$I(x, y) = s_1^2 + s_2^2 + (s_1 s_2)_+ + (s_1 s_2)_- \quad (2.5)$$

where

$$s_1^2 = \tfrac{1}{2} a_1^2(x, y) \langle \{1 + \cos 2[\omega t + \phi_1(x, y)]\} \rangle_t$$
$$s_2^2 = \tfrac{1}{2} a_2^2(x, y) \langle \{1 + \cos 2[\omega t + \phi_2(x, y)]\} \rangle_t$$
$$(s_1 s_2)_+ = \tfrac{1}{2} a_1(x, y) a_2(x, y) \langle \cos[2\omega t + \phi_1(x, y) + \phi_2(x, y)] \rangle_t$$
$$(s_1 s_2)_- = \tfrac{1}{2} a_1(x, y) a_2(x, y) \langle \cos[\phi_1(x, y) - \phi_2(x, y)] \rangle_t$$

This equation clearly shows why the temporal frequency term is so easily dropped. An oscillating function such as cos ωt averages to zero if many cycles are considered. Hence, Eq. (2.5) reduces to

$$I(x, y) = \tfrac{1}{2}\{a_1^2(x, y) + a_2^2(x, y) + a_1(x, y)a_2(x, y) \cos[\phi_1(x, y) - \phi_2(x, y)]\}$$
$$(2.6)$$

Note that we have succeeded in preserving the phase terms of both beams even though the recording was done on a phase insensitive medium. This, of course, is the "secret" of holography. For us to record, for example, a spherical wave on film we must record the phase, since a photograph of amplitude alone would show an almost uniform exposure which contains little information. An interferogram, however, shows a Fresnel pattern as discussed in Chapter 1.

So it seems that the preservation of phase is the overriding property of a hologram. It is a remarkable accomplishment in the field of optics, although as we saw, it is not new. In the long wavelength electro-magnetic region and in acoustics, the detection and preservation of phase is by no means startling. The reason for this is that detectors capable of measuring amplitude oscillations at the radian frequencies of the radiation are readily available. One need not use the indirect method of measuring phase as we did with optical radiation.

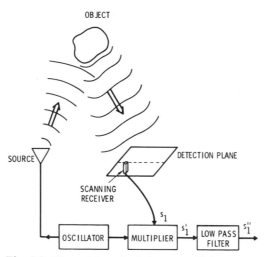

Fig. 2.2. The making of a hologram with long wavelengths can be done without a physical reference beam, as shown here. The local oscillator supplies a reference to a phase detector (multiplier and low-pass filter) while the receiver scans a plane.

For example, a common practice in communications is to use the system shown in Fig. 2.2. A physically small receiver is scanned over a plane. The received signal is multiplied by the oscillator signal and sent through a low-pass filter to yield the desired result. The mathematical procedure is as follows.

The received signal is

$$s_1(x, y) = a_1(x, y) \cos[\omega t + \phi_1(x, y)] \qquad (2.7)$$

After multiplication by $a_2 \cos(\omega t + \phi_2)$, this becomes

$$s'(x, y) = \frac{a_2 a_1(x, y)}{2} \{\cos[2\omega t + \phi_1(x, y) + \phi_2] + \cos[\phi_1(x, y) - \phi_2]\}$$
$$(2.8)$$

After low-pass filtering we are left with

$$s''(x, y) = \frac{a_2 a_1(x, y)}{2} \cos[\phi_1(x, y) - \phi_2] \qquad (2.9)$$

Comparing Eqs. (2.9) and (2.7) we see that we have succeeded in obtaining the same information without the attendant extraneous terms a_1^2 and a_2^2.

In addition to the indirect interferometric method, then, we can record a hologram by the direct detection method if long wavelength radiation is used. We will describe both methods in detail in succeeding chapters.

2.2. RECONSTRUCTION PROCESS

In Chapter 1 we noted that the wave front stored on the interferogram could be reconstructed by the process of diffraction of a plane wave. Here we generalize by using an arbitrary wave which can be expressed as

$$s_3(x, y) = a_3(x, y) \cos[\omega t + \phi_3(x, y)] \qquad (2.10)$$

Suppose we have exposed a photographic plate to the intensity shown in Eq. (2.6) and that after development it possesses a transmittance proportional to that intensity. Then the wave s_3, after passing through the plate is modified by it as

$$KI(x, y)s_3(x, y) = \frac{K}{2} \{a_1^2(x, y) + a_2^2(x, y)\}a_3(x, y) \cos[\omega t + \phi_3(x, y)]$$
$$+ \frac{K}{4} a_1(x, y)a_2(x, y)a_3(x, y)\{\cos[\omega t + \phi_1(x, y)$$
$$- \phi_2(x, y) + \phi_3(x, y)] + \cos[\omega t - \phi_1(x, y)$$
$$+ \phi_2(x, y) + \phi_3(x, y)]\} \qquad (2.11)$$

where K is constant.

If we consider the terms one at a time we note that the first term represents the reconstruction wave modified in amplitude. The second term, assuming that s_3 is a replica of s_2, becomes

$$\frac{K}{4} a_2^2(x, y)a_1(x, y) \cos[\omega t + \phi_1(x, y)] \qquad (2.12)$$

which we recognize as the original wave, s_1, with a modified amplitude. If s_2 happens to be a plane wave, a_2 is a constant, and a perfect reconstruction is obtained. With the same reconstruction wave, the third term of Eq. (2.11) becomes

$$\frac{K}{4} a_2^2(x, y)a_1(x, y) \cos[\omega t - \phi_1(x, y) + 2\phi_2(x, y)] \qquad (2.13)$$

This is the extraneous term we referred to earlier.

There are several more options available. For example, we could let s_3 be a replica of s_2 with opposite curvature; that is,

$$s_3(x, y) = a_2(x, y) \cos[\omega t - \phi_2(x, y)]$$

Then the second and third terms of Eq. (2.11) become

$$\frac{K}{4} a_2^2(x, y)a_1(x, y) \cos[\omega t + \phi_1(x, y) - 2\phi_2(x, y)]$$

and

$$\frac{K}{4} a_2^2(x, y)a_1(x, y) \cos[\omega t - \phi_1(x, y)] \qquad (2.14)$$

Here we see that the last term becomes a reconstruction of s_1 with opposite curvature and the second term is the extraneous image. Another possible form for s_3 is s_1, giving reconstructed terms

$$\frac{K}{4} a_1^2(x, y)a_2(x, y) \cos[\omega t + 2\phi_1(x, y) - \phi_2(x, y)]$$

and

$$\frac{K}{4} a_1^2(x, y)a_2(x, y) \cos[\omega t + \phi_2(x, y)] \qquad (2.15)$$

Thus, we have reconstructed s_2 and demonstrated the complete ambivalence of the interferometer. While recording the wave front of one arm, we have also recorded the other. Hence, either one may be reconstructed by diffracting the other beam through the interferogram. Additionally, by using the conjugate of one of the interfering beams as the diffracting beam, we can reconstruct the conjugate of the other.

If we are dealing with long wavelength radiation, we can use the direct method of detection to obtain the term shown in Eq. (2.9). However, to use this function to actually reconstruct a wave it must be written onto a medium capable of diffracting light. An alternative solution is to do the reconstruction by computer calculation, but then we still need to convert to a pictorial output.

To write the function [Eq. (2.9)] on something like film, we must introduce a bias term which makes the whole expression positive. This can be done by using the signal to modulate a light source about an appropriate bias intensity. This light source is photographed as it is scanned in synchronism with the receiver. The resulting photograph constitutes a hologram,

$$I(x, y) = I_b + \frac{a_2 a_1(x, y)}{2} \cos[\phi_1(x, y) - \phi_2] \qquad (2.15)$$

where

$$I_b \geq \frac{| \, a_2 a_1(x, y) \, |_{max}}{2}$$

We see that the only difference between this hologram and the interferometric one [Eq. (2.6)] is that the bias term is a constant, whereas in the former case it is space variant. The resulting diffraction effects are the same with the exception of the undiffracted term.

2.3. GABOR HOLOGRAPHY

We now consider Gabor's original method which is pictured in Fig. 2.3 [1]. The object is assumed to be rather transparent with the result that the transmitted wave may be considered to be made up of a large plane component and a small scattered component. Thus we can write Eq. (2.6) as

$$I(x, y) = \tfrac{1}{2}\{A^2 + a_2^2(x, y) + Aa_2(x, y) \cos \phi_2(x, y)\} \qquad (2.16)$$

where $A \cos \omega t$ represents the plane wave component and $a_2(x, y) \cos[\omega t + \phi_2(x, y)]$ the scattered component, $A \gg a_2$. When the film is developed and illuminated by another plane wave of uniform amplitude, $B \cos \omega t$, we obtain the result

$$\frac{K}{2} \{A^2 + a_2^2(x, y)\} B \cos \omega t + \frac{K}{4} ABa_2(x, y) \cos[\omega t - \phi_2(x, y)]$$

$$+ \frac{K}{4} ABa_2(x, y) \cos[\omega t + \phi_2(x, y)] \qquad (2.17)$$

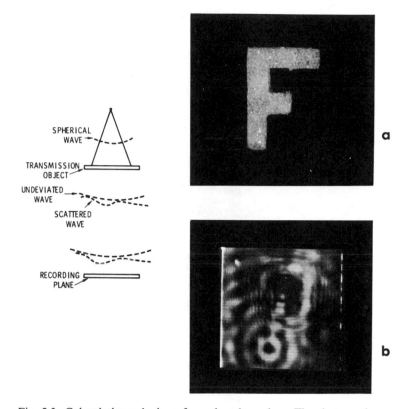

Fig. 2.3. Gabor holography is performed as shown here. The photographs show the object, (a) which consists of a small styrofoam letter F, and the resulting hologram. (b) The hologram consists of an intensity scan over a plane; the letter is 1 cm in height and the illumination spherical rather than plane as described in the text.

This is the mathematical equivalent of our physical illustration in Fig. 1.6. The last two terms are replicas of the scattered wave and its conjugate. Note that there is nothing to distinguish them except for the opposite sign on the phase term. Hence, they occupy the same volume of space as illustrated in Fig. 2.4. In addition, we have the first term which also propagates in the same direction. We can ignore the space variant portion $a_2{}^2(x, y)$ of the first term since $A^2 \gg a_2{}^2$. Thus, if we place a photographic plate at the position of the conjugate image we obtain a picture of the object added to a uniform background due to the true image wave propagated over the equivalent distance $2z_1$. If the object consists of a few dense points in a transparent field and if the distance $2z_1$ is quite large, the result will be

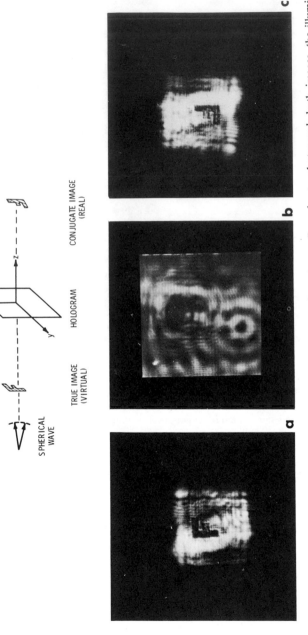

Fig. 2.4. The Gabor hologram recorded as in Fig. 2.3 is reconstructed as shown. In order to photograph both images, the illuminating wave was made to be converging so that both are real. Note that a rather large background obstructs the image; (a) is the true image, (b) the hologram, and (c) the conjugate image.

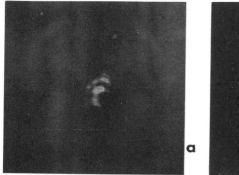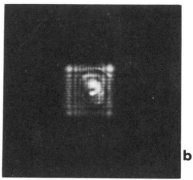

Fig. 2.5. This hologram (a) was made as shown in Fig. 2.3 with the exception of the object, which now consists of a sheet of styrofoam with the letter F cut out of it. The size of the letter and all other parameters are the same. Note that no image (b) is visible since practically no background penetrates the object.

quite good since the true image will be so badly out of focus that it merely contributes to the background. This is the reason that this type of holography has found use in particle sizing and distribution studies [4].

If we remove the restriction $A^2 \gg a_2^2$, we find that the plane wave component [first term of Eq. (2.17)] is not of uniform amplitude. This, in effect, throws a patterned shadow over the conjugate image, often obscuring it completely. For this reason, Gabor holography has found use only if the object is mostly transparent, thereby fulfilling the requirement $A^2 \gg a_2^2$. A reconstruction of a hologram of this type is shown in Fig. 2.5.

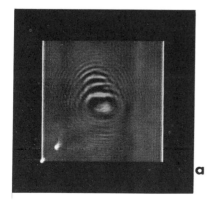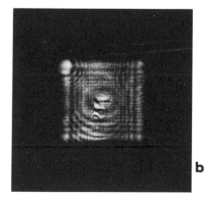

Fig. 2.6. In this figure all parameters are the same as in Fig. 2.5. Now, however, a coherent electronic reference is used. Note that the image (b) is now as good as that obtained in Fig. 2.4 even though no background wave is present.

It is interesting to note that for long wavelength radiation a Gabor hologram can be generated with a constant amplitude plane wave component without the aforementioned restriction, as shown by Eq. (2.15). A hologram of this type, and its reconstruction is shown in Fig. 2.6. The only remaining disadvantage for Gabor holography at long wavelengths is the interfering effect of the twin image. A number of techniques have been proposed in the past to overcome this problem without much success [5]. It was not until Leith and Upatnieks introduced the offset reference beam that a truly satisfactory method of separating the twin images was found [6].

2.4. LEITH–UPATNIEKS HOLOGRAPHY

An important innovation, introduced by E. N. Leith and J. Upatnieks, was to arrange the interfering beams in such a way that their average directions of propagation were not collinear as in Gabor holography. In this way, the interferogram became a linear grating structure upon which was superimposed a spatial deviation, proportional to the departures from a plane wave that might exist on either or both beams. This situation was simply demonstrated in Fig. 2.1. We now demonstrate these statements mathematically.

A plane wave intercepting a plane at an angle ϕ can be expressed as a function of coordinates x, y on that plane as

$$s(x, y) = a(x, y) \cos[\omega t + \alpha x] \qquad (2.18)$$

where

$$\alpha = \frac{\sin \phi}{\lambda}$$

and we have assumed that the direction of propagation is in the (x, z) plane.

We can, therefore, express the amplitudes of beams 1 and 2 in terms of a plane front and a phase deviation from that plane. That is,

$$s_1(x, y) = a_1(x, y) \cos[\omega t + \alpha_1 x + \Delta\phi_1(x, y)]$$
$$s_2(x, y) = a_2(x, y) \cos[\omega t + \alpha_2 x + \Delta\phi_2(x, y)]$$

where

$$\alpha x + \Delta\phi(x, y) = \phi(x, y)$$

and

$$\alpha_1 = \frac{\sin \phi_1}{\lambda}, \qquad \alpha_2 = \frac{\sin \phi_2}{\lambda}$$

Similarly, the reconstruction beam can be written

$$s_3(x, y) = a_3(x, y) \cos[\omega t + \alpha_3 x + \Delta\phi_3(x, y)]$$

The diffracted wave then becomes

$$\frac{K}{2} \{[a_1^2(x, y) + a_2^2(x, y)]a_3(x, y) \cos[\omega t + \alpha_3 x + \Delta\phi_3(x, y)]\}$$
$$+ \frac{K}{4} a_1(x, y)a_2(x, y)a_3(x, y)\{\cos[\omega t + (\alpha_1 - \alpha_2 + \alpha_3)x$$
$$+ \Delta\phi_1(x, y) - \Delta\phi_2(x, y) + \Delta\phi_3(x, y)] + \cos[\omega t + (\alpha_2 - \alpha_1 + \alpha_3)x$$
$$- \Delta\phi_1(x, y) + \Delta\phi_2(x, y) + \Delta\phi_3(x, y)]\} \tag{2.19}$$

This expression is equivalent to Eq. (2.11). If we now assume that $s_3 = s_2$ as we did previously, we find the last two terms to be

$$\frac{K}{4} a_2^2(x, y)a_1(x, y) \cos[\omega t + \alpha_1 x + \Delta\phi_1(x, y)] \tag{2.20}$$

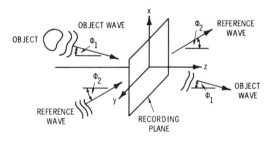

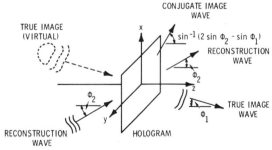

Fig. 2.7. The physical relationship between object, reference, reconstruction and image waves for Leith–Upatnieks holography is shown here. In this case we are illuminating the hologram so as to obtain the true image. Note the complete physical separation of the various waves.

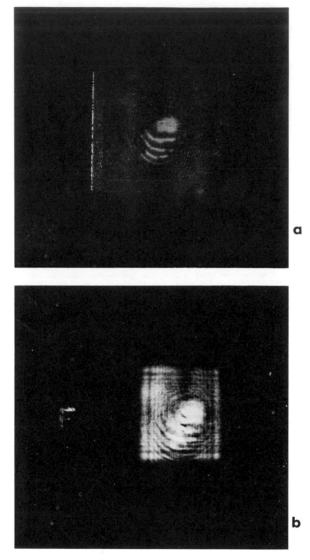

Fig. 2.8. This hologram (a) and its reconstructed true image (b) demonstrate the power of the Leith–Upatnieks technique. The image is completely separated from the zero-order light and the out-of-focus conjugate image.

and

$$\frac{K}{4} a_2{}^2(x, y)a_1(x, y) \cos[\omega t + (2\alpha_2 - \alpha_1)x - \Delta\phi_1(x, y) + 2\Delta\phi_2(x, y)] \quad (2.21)$$

These expressions are the equivalents to Eqs. (2.12) and (2.14), but are in a form suited to illustrating the separation of the diffracted beams. Equation (2.20) represents a replica of $s_1(x, y)$ propagating at the angle ϕ_1. Equation (2.21) represents the extraneous beam propagating in the direction

$$\sin^{-1}[2 \sin \phi_2 - \sin \phi_1] \quad (2.22)$$

which for small angles can be written as $2\phi_2 - \phi_1$.

The complete Leith–Upatnieks holography system is shown in Fig. 2.7 where we have assumed $s_3(x, y) = s_2(x, y)$. We see that we can observe the true image with complete freedom from extraneous light. Figure 2.8 is a reconstruction of such an acoustical hologram.

2.5. HOLOGRAM CLASSIFICATION

This is a convenient place to briefly consider various classifications that have been given to holograms made under certain specific conditions. These classifications have been discussed in great detail by DeVelis and Reynolds so we will only mention them [7].

2.5.1. Fresnel Holograms

A Fresnel hologram is defined as the interference pattern between two collinear radiation beams, where the pattern is recorded in the Fresnel zone of the object generating one of the beams [8]. This simply means that if one beam is spherical and the other plane, we obtain a Fresnel pattern. Since the beams are collinear, we have redefined the Gabor hologram with its attendant disadvantages. Images from this type of hologram using different objects were shown in Figs. 2.4 and 2.5.

2.5.2. Fraunhofer Holograms

A Fraunhofer hologram is defined as the interference pattern between two collinear radiation beams, where the pattern is recorded in the far field or Fraunhofer zone of the object generating one of the beams [9]. Using our example of a spherical and plane beam, this means that the hologram is made so far away from the center of curvature of the spherical beam that it appears plane over the extent of the hologram. The advantage of this type of hologram over the Fresnel hologram is that the twin image is so far away from the image we wish to photograph that it does not contribute much noise or interference.

2.5.3. Side-Band Fresnel or Fraunhofer Holograms

We previously discussed the essential points of this class of holography when we introduced the idea of noncollinear beams as suggested by Leith and Upatnieks. Both Fresnel and Fraunhofer holograms can be improved by making the two interfering beams noncollinear, the improvement being the complete separation of the twin images. There are certain advantages and disadvantages of side-band Fresnel over side-band Fraunhofer holograms related to the resolution of the recording medium [10]. An example of the improvement offered by this method was shown in Fig. 2.8.

2.5.4. Fourier Transform Holograms

A Fourier transform hologram is defined as the interference pattern of a plane wave with the spatial Fourier transform of the object. The spatial Fourier transform of an object may be obtained in several ways. The simplest way is to move the observation plane into the far zone of the object. In the far zone, each point on the object manifests itself as a plane wave propagating in a unique direction. When this is interfered with a plane reference wave, the result is a sum of linear interference lines of different spacing and orientation. This, after all, is what a Fourier transform is; a decomposition of an arbitrary function into a sum of sinusoidal functions of different frequencies.

Another way of obtaining a spatial Fourier transform is to artificially remove the object into the far zone. This is simply done by placing it in the front focal plane of a lens. In this position, the lens transforms each point on the object to a plane wave, which is equivalent to moving it into the far field.

Since such a hologram will reconstruct a Fourier transform rather than an image, it is necessary to use a lens in the diffracted field. In a sense, one may think of the lens as performing an inverse Fourier transform to recover the image. An example of such a hologram and its reconstructed image is shown in Fig. 2.9.

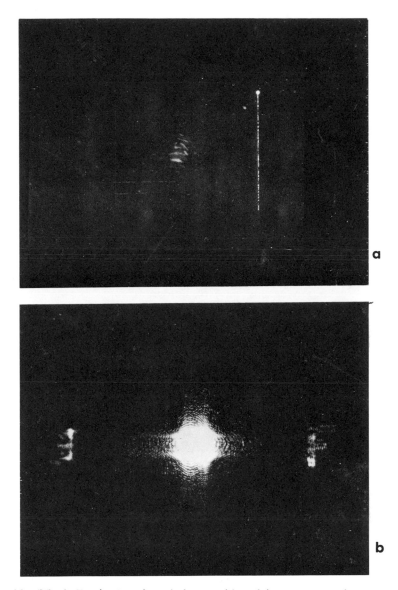

Fig. 2.9. A Fourier transform hologram (a) and its reconstructed true
image (b) are shown in these photographs. Note that the dominant structure of the hologram is now linear rather than circular as for the Fresnel
hologram. Both images should be in focus, but due to imperfect acoustic
lens placement this is not so.

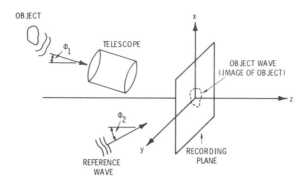

Fig. 2.10. The focused-image hologram is made by imaging the object into the recording plane as shown here.

2.5.5. Focused-Image Holograms

This is one classification of holography that has not been discussed in detail. A definition of this class of hologram might be the interference pattern between two noncollinear beams of radiation recorded in the near zone of the object. Physically, this is possible only if the hologram is recorded by casting an image of the object into the hologram plane as shown in Fig. 2.10. Naturally, when such a hologram is reilluminated the image will appear in the hologram aperture. This class of holography is important in optical contouring [11] and in liquid-surface acoustical holography as will be shown in following chapters. An example of such a hologram and its reconstructed image appears in Fig. 2.11.

2.6. IMAGE LOCATION

Up to this point in our discussions we have obtained only qualitative results. What happens if we use different wavelengths or propagation media for recording and illuminating the hologram, or if the illuminating beam is not an exact replica of the reference beam? To answer these questions we must perform a quantitative analysis based upon the very general arrangement shown in Fig. 2.12.

In this analysis we assume that the primary radiation source is a perfect point. Figure 2.12 shows the general recording arrangement, where for simplicity we have shown only two dimensions. We have three general structures, two of which are identically illuminated by the primary point source. These two are the reference and illumination structures. The radia-

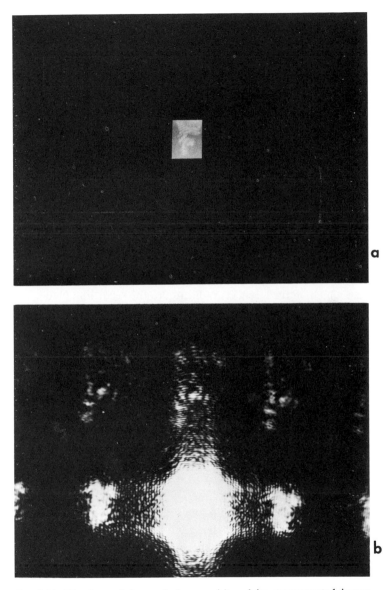

Fig. 2.11. The focused-image hologram (a) and its reconstructed image
(b) appear in these photographs. Note that the hologram information is
stored on a very small part of the aperture; namely the focused F. Hence
we obtain only a weak image.

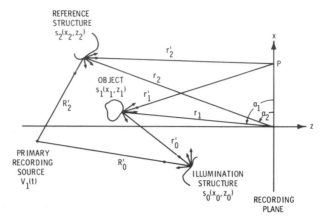

Fig. 2.12. This drawing establishes the nomenclature for the recording step for our general analysis. We allow complete generality by considering the coherence of the source, and arbitrary illumination and reference structures.

tion reflected from or transmitted by the illumination structure illuminates the object which, in turn, reflects radiation to the recording plane. The radiation transmitted or reflected by the reference structure also illuminates the recording plane. We assume that the complex amplitude of the radiation emanating from the primary source may be written $V_1(t)$.

The radiation reaching a point P at the hologram plane from the illuminating structure via the object is

$$U_1 = A_1 \int s_0(x_0, z_0) s_1(x_1, z_1) V_1\left(t - \frac{R_0' + r_0' + r_1'}{c_1}\right) dx_0 \, dz_0 \, dx_1 \, dz_1 \quad (2.23)$$

where c_1 is the velocity of propagation, $s_0(x_0, z_0)$, $s_1(x_1, z_1)$ are the descriptions of the radiation field at the surface of the illumination structure and the object, respectively, and $(R_0' + r_0' + r_1')/c_1$ is the transit time of the radiation from the source to the point P via the path indicated in Fig. 2.12. Here we have assumed that the change in $r_0' + r_1' + R_0'$ over the illumination and source structure is small enough so that the $1/r$ amplitude dependence can be considered constant. This is expressed in Eq. (2.23) by the term A_1. This assumption prevents this analysis from being correct in the near field. Similarly, the radiation reaching the point P from the reference structure is

$$U_2 = A_2 \int s_2(x_2, z_2) V_1\left(t - \frac{R_2' + r_2'}{c_1}\right) dx_2 \, dz_2 \quad (2.24)$$

The quantity recorded is the time average of the intensity expressed as

$$I = \langle |U_1 + U_2|^2 \rangle_t = \langle |U_1|^2 \rangle_t + \langle |U_2|^2 \rangle_t + 2Re\langle U_1 U_2^* \rangle_t \quad (2.25)$$

The first two terms may be identified as the intensity at P due to the object and reference beams, respectively. The last term becomes

$$2Re\langle U_1 U_2^* \rangle_t = 2Re \int s_1(x_1, z_1)s_0(x_0, z_0)s_2^*(x_2, z_2) \left\langle V_1\left(t - \frac{R_0' + r_0' + r_1'}{c_1}\right) \right.$$

$$\left. \times V_1^*\left(t - \frac{R_2' + r_2'}{c_1}\right) \right\rangle_t dx_1\, dz_1\, dx_0\, dz_0\, dx_2\, dz_2 \quad (2.26)$$

The term $\langle V_1(t_1)V_1^*(t_2) \rangle_t = \Gamma_{11}(t_1, t_2)$ is known as the correlation function of the field at the two times, t_1 and t_2 [12]. For long-term time averaging, such as that occurring in the photographic process, the correlation function depends only on the time difference $t_1 - t_2$. Therefore,

$$\left\langle V_1\left(t - \frac{R_0' + r_0' + r_1'}{c_1}\right) V_1^*\left(t - \frac{R_2' + r_2'}{c_1}\right) \right\rangle_t$$

$$= \Gamma_{11}\left[\frac{(R_2' + r_2') - (R_0' + r_0' + r_1')}{c_1} \right] \quad (2.27)$$

Equation (2.25) becomes

$$I = I_1 + I_2 + A_1 A_2^* \int s_1(x_1, z_1)s_0(x_0, z_0)s_2^*(x_2, z_2)$$

$$\times \Gamma_{11}\left[\frac{(R_2' + r_2') - (R_0' + r_0' + r_1')}{c_1} \right] dx_1\, dz_1\, dx_0\, dz_0\, dx_2\, dz_2$$

$$+ A_1^* A_2 \int s_1^*(x_1, z_1)s_0^*(x_0, z_0)s_2(x_2, z_2)$$

$$\times \Gamma_{11}\left[\frac{(R_0' + r_0' + r_1') - (R_2' + r_2')}{c_1} \right] dx_1\, dz_1\, dx_0\, dz_0\, dx_2\, dz_2 \quad (2.28)$$

We assume that the optical density on the developed film is proportional to I. Suppose we illuminate the hologram as shown in Fig. 2.13. Another primary point source strikes a reconstruction structure which, in turn, illuminates the hologram. We look at an arbitrary point Q and determine the complex amplitude of the light at that point. The complex amplitude of the light falling on the hologram at P is proportional to

$$U_a = A_a \int s_a(x_a, z_a)V_2\left(t - \frac{R_a' + r_a'}{c_2}\right) dx_a\, dz_a \quad (2.29)$$

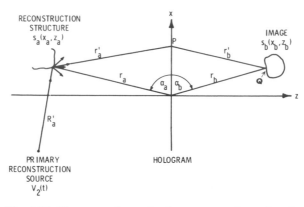

Fig. 2.13. The nomenclature for the reconstruction or imaging
step is established in this figure.

where c_2 is the velocity of propagation of the light. The amplitude of the
light emerging from the hologram is the multiplication of Eq. (2.29) by
Eq. (2.28). The amplitude of the light reaching Q is proportional to

$$
\begin{aligned}
s_b(x_b, z_b) = & \int [I_1(x) + I_2(x)] s_a(x_a, z_a) V_2\left(t - \frac{R_a' + r_a' + r_b'}{c_2}\right) dx\, dx_a\, dz_a \\
& + A_1 A_2^* \int s_1(x_1, z_1) s_0(x_0, z_0) s_2^*(s_2, z_2) s_a(x_a, z_a) \\
& \times \Gamma_{11}\left[\frac{(R_2' + r_2') - (R_0' + r_0' + r_1')}{c_1}\right] V_2\left(t - \frac{R_a' + r_a' + r_b'}{c_2}\right) \\
& \times dx_1\, dz_1\, dx_0\, dz_0\, dx_2\, dz_2\, dx_a\, dz_a\, dx + A_1^* A_2 \int s_1^*(x_1, z_1) \\
& \times s_0^*(x_0, z_0) s_2(x_2, z_2) s_a(x_a, z_a) \Gamma_{11}\left[\frac{(R_0' + r_0' + r_1') - (R_2' + r_2')}{c_1}\right] \\
& \times V_2\left(t - \frac{R_a' + r_a' + r_b'}{c_2}\right) dx_1\, dz_1\, dx_0\, dz_0\, dx_2\, dz_2\, dx_a\, dz_a\, dx \\
& \hspace{10cm} (2.30)
\end{aligned}
$$

By the Weiner–Kintchine theorem, we may replace the correlation
function by a power spectral density [13]. That is,

$$
\Gamma_{11}(\tau) = \frac{1}{2\pi} \int_{-\infty}^{\infty} \phi_{11}(\omega_1) \exp(i\omega_1 \tau)\, d\omega_1 \tag{2.31}
$$

where $\phi_{11}(\omega_1)$ is the power spectral density of $V_1(t)$.

Let us also replace $V_2(t)$ by its spectral description,

$$
V_2(t) = \frac{1}{2\pi} \int_{-\infty}^{\infty} v_2(\omega_2) \exp(i\omega_2 t)\, d\omega_2 \tag{2.32}
$$

where $v_2(\omega_2)$ is the Fourier transform of $V_2(t)$. Before we rewrite Eq. (2.30) with Eqs. (2.31) and (2.32) substituted, let us look at each term and work only with the most significant one. The first term amplitude modulates the beam, but does not diffract it. Therefore, this term does not contribute to the image. We can consider the last two terms simultaneously since they are complex conjugates. Let us use the first term which will be seen to be the true image term. It becomes

$$\frac{A_1 A_2^*}{(2\pi)^2} \int s_1(x_1, z_1) s_0(x_0, z_0) s_2^*(x_2, z_2) s_a(x_a, z_a) v_2(\omega_2) \phi_{11}(\omega_1)$$

$$\times \exp\left\{+i\omega_2\left[t - \frac{R_a' + r_a' + r_b' + k_1/k_2(R_0' + r_0' + r_1' - R_2' - r_2')}{c_2}\right]\right\}$$

$$\times dx_1 \, dz_1 \, dx_0 \, dz_0 \, dx_2 \, dz_2 \, dx_a \, dz_a \, d\omega_2 \, d\omega_1 \, dx \qquad (2.33)$$

where $k_1 = \omega_1/c_1 = 2\pi/\lambda_1$ and $k_2 = \omega_2/c_2 = 2\pi/\lambda_2$.

Suppose we regroup the terms of Eq. (2.33) in the following way [14]:

$$\frac{A_1 A_2^*}{(2\pi)^2} \int \left[\int\left[\int s_1(x_1, z_1) f_1 f_2 \, dx_1 \, dz_1 \, dx\right] s_0(x_0, z_0) s_2^*(x_2, z_2) s_a(x_a, z_a) v_2(\omega_2) \phi_{11}(\omega_1)\right.$$

$$\times \exp[i\omega_2 t] \, dx_0 \, dz_0 \, dx_2 \, dz_2 \, dx_a \, dz_a \, d\omega_1 \, d\omega_2 \qquad (2.34)$$

where

$$f_1 = \exp\{-i[k_2(R_a' + r_a') + k_1(R_0' + r_0' + r_1' - R_2' - r_2')]\} \qquad (2.35)$$

and

$$f_2 = \exp(-ik_2 r_b') \qquad (2.36)$$

We wish to show that the integral in the square bracket of Eq. (2.34) is an image of the object. If we assume that it is, it must be expressible as

$$s_b(x_b, z_b) = s_1(x_b, z_b) \exp[-i\theta(x_b, z_b)] \qquad (2.37)$$

where θ is an unknown phase term introduced by the process. This result can be obtained only if

$$f_1 = \exp\{i[k_2 r_b'(x - x_1, z_1) - \theta(x_1, z_1)]\} \qquad (2.38)$$

If this expression is substituted into the bracketed term of Eq. (2.34), the result is

$$s = \int\left\{\int s_1(x_1, z_1) \exp[-i\theta(x_1, z_1)] \times \exp[ik_2 r_b'(x - x_1, z_1)] \, dx_1 \, dz_1\right\}$$

$$\times \{\exp[-ik_2 r_b'(x - x_b, z_b)]\} \, dx \qquad (2.39)$$

Equation (2.39) may be interpreted as a diffraction of s_1 followed by an inverse diffraction into an image of s_1. Of course, the $1/r$ dependence of the diffraction integral has been neglected under the assumption that it is nearly constant over the domain of integration.

By equating Eq. (2.38) to Eq. (2.35), we obtain the usual solutions that perfect reconstruction occurs only for $k_1 = k_2$, $r_a' = r_2'$ in which case $z_b = -z_1$ and $x_b = x_1$ [14]. In other words, the image duplicates the object in position and size.

We may expand the distances r_1', r_2', r_a', r_b' about the origin to obtain the solutions for the Gaussian image location [15]. As an example, we expand r_1' as follows:

$$r_1' = [(x - x_1)^2 + (z_1)^2]^{\frac{1}{2}} = [(x_1^2 + z_1^2) + x(x - 2x_1)]^{\frac{1}{2}} \qquad (2.40)$$

From Fig. 2.12 we see that $(x_1^2 + z_1^2)$ may be designated as r_1^2. Then, by expanding Eq. (2.40) about r_1 in a binomial series and retaining only the first two terms we obtain

$$r_1' = r_1 + \frac{x(x - 2x_1)}{2r_1} \qquad (2.41)$$

If we do this for r_2', r_a', r_b' as well, and substitute into Eqs. (2.35) and (2.36), we have

$$f_1 = \exp\left\{-i\left[k_2\left(R_a' + r_a + \frac{x(x - 2x_a)}{2r_a}\right)\right.\right.$$
$$\left.\left. + k_1\left(R_0' - R_2' + r_0' + r_1 + \frac{x(x - 2x_1)}{2r_1} - r_2 - \frac{x(x - 2x_2)}{2r_2}\right)\right]\right\} \qquad (2.42)$$

and

$$f_2 = \exp\left\{-ik_2\left[r_b + \left(\frac{x(x - 2x_b)}{2r_b}\right)\right]\right\} \qquad (2.43)$$

For imaging, f_1 must equal

$$f_1 = \exp\left\{i\left[k_2\left(r_b + \frac{x(x - 2x_b)}{2r_b}\right) - \theta\right]\right\} \qquad (2.44)$$

This is true for

$$\cos \alpha_b = -\frac{k_1}{k_2}(\cos \alpha_1 - \cos \alpha_2) - \cos \alpha_a \qquad (2.45)$$

$$1/r_b = -\frac{k_1}{k_2}(1/r_1 - 1/r_2) - 1/r_a \qquad (2.46)$$

and

$$\theta = -k_1(R_0' + r_0' + r_1 - R_2' - r_2) + k_2(R_a' + r_a + r_b) \quad (2.47)$$

where $\cos \alpha_b = x_b/r_b$, $\cos \alpha_1 = x_1/r_1$, $\cos \alpha_a = x_a/r_a$. A negative value for r_b indicates a point in the left half plane (a virtual point). Equation (2.34) may now be written as

$$\frac{A_1 A_2^*}{(2\pi)^2} \int s_1(x_b, z_b)$$

$$\times \exp\{-i[k_1(R_0' + r_0' + r_1 - R_2' - r_2) + k_2(R_a' + r_a + r_b)]\}$$

$$\times \exp[i\omega_2 t]s_0(x_0, z_0)s_2^*(x_2, z_2)s_a(x_a, z_a)\phi_{11}(\omega_1)v_2(\omega_2)$$

$$\times dx_0 \, dz_0 \, dx_2 \, dz_2 \, dx_a \, dz_a \, d\omega_1 \, d\omega_2 \quad (2.48)$$

From this equation we see that the image is modified by the illumination, reference, and reconstruction structures and the frequency characteristic of the primary sources V_1 and V_2. We note from Eqs. (2.45) and (2.46) that $s_1(x_b, z_b)$ is dependent upon s_2 and s_a through the coordinates x_2, z_2, x_a, z_a and upon the frequency spread of ω_1 and ω_2. To get an image, then, we must have $s_a(x_a, z_a)$, $s_2(x_2, z_2)$ related in such a way that

$$\frac{k_1}{k_2 r_2} - \frac{1}{r_a} = \text{const} \quad (2.49)$$

and

$$\frac{k_1}{k_2} \cos \alpha_2 - \cos \alpha_a = \text{const} \quad (2.50)$$

and ω_1, ω_2 constant. This can be done in several ways, the simplest of which is to use single point structures. Then r_2, r_a, α_2, α_a are constants and together with monochromatic radiation, the problem is solved. For the general case of a complex reference structure, the only way imaging can be achieved is if the reconstruction structure is an exact replica of the reference structure, with the scale factor according to Eqs. (2.49) and (2.50) with suitably chosen constants. The simplest example, of course, is to choose the constants to be zero, in which case

$$r_a = \frac{k_2}{k_1} r_2$$

$$\cos \alpha_a = \frac{k_1}{k_2} \cos \alpha_2$$

with the result that

$$r_b = -\frac{k_2}{k_1} r_1, \qquad \cos \alpha_b = -\frac{k_1}{k_2} \cos \alpha_1$$

When Eqs. (2.49) and (2.50) are met, we have the result that the image amplitude is

$$s_b(x_b, z_b) = K \Big\{ s_1(x_b, z_b) \exp\Big[-ik_2\Big(r_b - \frac{k_1}{k_2} r_1\Big)\Big] \Big\}$$

$$\times \Big\{ \int s_0(x_0, z_0) \exp[ik_1(R_0' + r_0')] \, dx_0 \, dz_0 \Big\}$$

$$\times \Big\{ \int s_2(x_2, z_2) \exp[ik_1 R_2'] \, dx_2 \, dz_2 \Big\}^*$$

$$\times \Big\{ \int s_a(x_a, z_a) \exp[-ik_2 R_a'] \, dx_a \, dz_a \Big\} \qquad (2.51)$$

where K is a constant. This expression shows that an image has been obtained. The image is a scaled replica of the original object with additional phase and amplitude shading. The first term of Eq. (2.51) represents the image, the second bracket represents the illumination on the object, the last two terms describe the effect of the reference and reconstruction structures. These last two terms correspond to the amplitude shading expressed in Eq. (2.12).

We have obtained expressions for the true image; the conjugate image is just as easily obtained, and is the complex conjugate of Eq. (2.51). The image location equations become

$$\frac{1}{r_b} = \frac{k_1}{k_2} \Big(\frac{1}{r_1} - \frac{1}{r_2}\Big) - \frac{1}{r_a} \qquad (2.52)$$

$$\cos \alpha_b = \frac{k_1}{k_2} (\cos \alpha_1 - \cos \alpha_2) - \cos \alpha_a \qquad (2.53)$$

In the mathematical manipulation described above, we chose not to introduce the additional complexity of allowing for an enlargement or reduction of the hologram itself. This is not difficult to do, however, and so we write the image location equations including this factor, m, and allowing for both true and conjugate images below

$$\frac{1}{r_b} = \pm \frac{k_1}{m^2 k_2} \Big(\frac{1}{r_1} - \frac{1}{r_2}\Big) - \frac{1}{r_a} \qquad (2.54)$$

$$\cos \alpha_b = \pm \frac{k_1}{mk_2} (\cos \alpha_1 - \cos \alpha_2) - \cos \alpha_a \qquad (2.55)$$

where the lower sign refers to the true image, the upper sign to the conjugate image.

2.7. MAGNIFICATION

Magnification expressions may easily be obtained by performing the indicated differentiations.

$$\text{Lateral magnification } M_l = \frac{\partial x_b}{\partial x_1} = \pm \frac{k_1}{mk_2} \frac{r_b}{r_1} \tag{2.56}$$

$$\text{Radial magnification } M_r = \frac{\partial r_b}{\partial r_1} = \pm \frac{k_1}{m^2 k_2} \left(\frac{r_b}{r_1}\right)^2 \tag{2.57}$$

$$\text{Angular magnification } M_\alpha = \frac{\partial \alpha_b}{\partial \alpha_1} = \pm \frac{k_1}{mk_2} \frac{\sin \alpha_1}{\sin \alpha_b} \tag{2.58}$$

Another magnification expression sometimes more practical than lateral is one which is in a plane perpendicular to the radius vector to the image point. This is defined as

$$\text{Normal magnification } M_n = \frac{r_b}{r_1} \frac{\partial \alpha_b}{\partial \alpha_1} = \pm \frac{k_1}{mk_2} \frac{r_b}{r_1} \frac{\sin \alpha_1}{\sin \alpha_b} \tag{2.59}$$

2.8. ABERRATIONS

As with any imaging system, the holographic image suffers from aberrations if the recording and reconstruction steps are not carried out under identical conditions. The aberrations are the same ones that lens systems are plagued with, and are calculated and classified in the same way.

We recall that in obtaining the image-location equations we expanded the distance terms in a binomial series and neglected coefficients of $x^2(x - 2x_p)^2$ and higher. This gave us a first approximation to the location of the image point. If we retain coefficients of $x^2(x - 2x_p)^2$ and neglect those of higher order, we can obtain the so-called third-order aberrations. Equations (2.42) and (2.43) become

$$f_1 = \exp\left\{-i\left[k_2\left(R_a' + r_a + \frac{x(x - 2x_a)}{2r_a} - \frac{x^2(x - 2x_a)^2}{8r_a^3}\right)\right.\right.$$
$$+ k_1\left(R_0' - R_2' + r_0' + r_1 + \frac{x(x - 2x_1)}{2r_1} - \frac{x^2(x - 2x_1)^2}{8r_1^3}\right.$$
$$\left.\left.- r_2 - \frac{x(x - 2x_2)}{2r_2} + \frac{x^2(x - 2x_2)^2}{8r_2^3}\right)\right]\right\} \tag{2.60}$$

and

$$f_2 = \exp\left\{-i\left[k_2\left(r_b + \frac{x(x - 2x_b)}{2r_b} - \frac{x^2(x - 2x_b)^2}{8r_b^3}\right)\right]\right\} \tag{2.61}$$

As before, for imaging f_1 must equal

$$f_1 = \exp\left\{i\left[k_2\left(r_b + \frac{x(x - 2x_b)}{2r_b} - \frac{x^2(x - 2x_b)^2}{8r_b^3}\right) - \theta\right]\right\} \quad (2.62)$$

Equating (2.60) and (2.62) we obtain the same image location equations but the expression for θ becomes

$$\theta = -k_1(R_0' + r_0' + r_1 - R_2' - r_2) + k_2(R_a' + r_a + r_b)$$

$$+ \frac{x^4 k_2}{8}\left[-\frac{k_1}{k_2}\left(\frac{1}{r_1^3} - \frac{1}{r_2^3}\right) - \frac{1}{r_a^3} - \frac{1}{r_b^3}\right]$$

$$+ \frac{x^3 k_2}{2}\left[+\frac{k_1}{k_2}\left(\frac{x_1}{r_1^3} - \frac{x_2}{r_2^3}\right) + \frac{x_a}{r_a^3} + \frac{x_b}{r_b^3}\right]$$

$$+ \frac{x^2 k_2}{2}\left[-\frac{k_1}{k_2}\left(\frac{x_1^2}{r_1^3} - \frac{x_2^2}{r_2^3}\right) - \frac{x_a^2}{r_a^3} - \frac{x_b^2}{r_b^3}\right] \quad (2.63)$$

Before we introduced the higher order terms in the binomial expansion, we had θ free of the variable x. Consequently, we were able to reduce the integral in Eq. (2.34). We can still proceed along the same lines but must now admit that errors exist. These errors have been given particular names. For instance, the coefficients of x^4, x^3, x^2 are given the respective names of spherical, coma, and astigmatism aberrations. If we had used three dimensions in our analysis, we would have obtained terms in y^4, y^3, y^2, and $(xy)^2$ as well. The complete aberration terms for both true and conjugate images are given below.

Spherical (coefficient of $(x^4 + y^4)$)

$$S = \pm \frac{k_1}{m^4 k_2}\left(\frac{1}{r_1^3} - \frac{1}{r_2^3}\right) + \frac{1}{r_a^3} + \frac{1}{r_b^3} \quad (2.64)$$

Coma (coefficients of x^3, y^3)

$$C_x = \pm \frac{k_1}{m^3 k_2}\left(\frac{x_1}{r_1^3} - \frac{x_2}{r_2^3}\right) + \frac{x_a}{r_a^3} + \frac{x_b}{r_b^3} \quad (2.65)$$

$$C_y = \pm \frac{k_1}{m^3 k_2}\left(\frac{y_1}{r_1^3} - \frac{y_2}{r_2^3}\right) + \frac{y_a}{r_a^3} + \frac{y_b}{r_b^3}$$

Astigmatism (coefficients of x^2, y^2)

$$A_x = \pm \frac{k_1}{m^2 k_2}\left(\frac{x_1^2}{r_1^3} - \frac{x_2^2}{r_2^3}\right) + \frac{x_a^2}{r_a^3} + \frac{x_b^2}{r_b^3} \quad (2.66)$$

$$A_y = \pm \frac{k_1}{m^2 k_2}\left(\frac{y_1^2}{r_1^3} - \frac{y_2^2}{r_2^3}\right) + \frac{y_a^2}{r_a^3} + \frac{y_b^2}{r_b^3}$$

Field curvature (coefficients of $(xy)^2$)

$$A_{xy} = \pm \frac{k_1}{m^2 k_2} \left(\frac{x_1 y_1}{r_1^{\,3}} - \frac{x_2 y_2}{r_2^{\,3}} \right) + \frac{x_a y_a}{r_a^{\,3}} + \frac{x_b y_b}{r_b^{\,3}} \qquad (2.67)$$

In conventional aberration theory an additional third order aberration exists; namely, distortion. This term describes the fact that an imaging system with magnification normally images a plane onto a sphere. Since we expanded all the distance terms, including the image distance r_b' about a radius, we have assumed that the image lies on the sphere of radius r_b with the result that this aberration does not exist. In reality, however, we would project the image onto a plane so that we would see distortion. We will not dwell upon this matter here since adequate analyses exist [15–17].

A study of Eqs. (2.64)–(2.67), together with Eqs. (2.54) and (2.55), reveals various conditions under which some or all of the aberrations disappear. All aberrations disappear when $k_1 = k_2$, $m = 1$, $r_a = r_2$, $\cos \alpha_a = \pm \cos \alpha_2$. This, of course, results in a perfect image with $r_b = \pm r_1$, $\cos \alpha_b = \pm \cos \alpha_1$, and magnification of unity. Another set of conditions yielding a nonaberrated image is $m = k_1/k_2$, $r_a = r_2 = \infty$, $\cos \alpha_a = \pm \cos \alpha_2$, which results in an image with the characteristics $r_b = \pm k_1 r_1/k_2$, $\cos \alpha_b = \pm \cos \alpha_1$, and $M_n = \pm k_1/k_2$, $M_r = \pm k_1/k_2$. Thus, for this special case, we can obtain undistorted and unaberrated magnification.

2.9. DISTORTION

Distortion as used in this section is different from distortion aberration mentioned in Section 2.8. We could have introduced distortion of the image in our discussion of magnification, since that is where it arises. However, in connection with scanned holography, which we discuss later in this book, we introduce a very important concept that allows one to overcome this type of distortion. Consequently, we feel this discussion is appropriate to include here.

If we consider the magnification equations, we see that radial magnification has, in general, a different magnitude than the lateral magnification. This is true of all imaging systems. A lens, for example, has the relationship

$$M_r = M_l^2 \qquad (2.68)$$

where we have used the paraxial approximation. This formula is known as Maxwell's elongation formula [18].

For the paraxial holographic imaging system, the equivalent Maxwell's elongation formula is

$$M_r = \pm \frac{k_2}{k_1} M_l^2 \tag{2.69}$$

which was obtained from Eqs. (2.56) and (2.57). For the more general case it is best to use the normal magnification relation to obtain

$$M_r = \pm \frac{k_2}{k_1} \left(\frac{\sin \alpha_b}{\sin \alpha_1} \right)^2 M_n^2 \tag{2.70}$$

We see from these expressions that this type of distortion of the image is the normal course of events for all imaging systems with holography suffering the additional factor k_1/k_2. We previously discussed the one exception to this rule in Section 2.8.

2.10. RESOLUTION

The concept of the "quality" of an imaging system has always been a subject for argument. The question is academic, since the definition of quality depends upon the intended use of the final image. In astronomy the problem might be to resolve a double star. In this case, the quality criterion is the ability to resolve two incoherent points closely spaced. A common radiographic problem is to measure the swelling of a radioactive fuel element, for which one would like to have an image with sharp edges. Consequently, the "best" image is one that produces sharp edges and the fidelity of the rest of the image is immaterial. Another criterion sometimes considered is that the mean-square-error between image and object be a minimum.

The difficulties mentioned above are compounded by the additional question of coherent versus incoherent illumination. For instance, the two-point resolution or Rayleigh criterion may be meaningless if coherent illumination is used. A general discussion of this whole problem is available in Goodman's book [19].

In view of the nebulousness regarding the definition of resolution, we use a definition that relates to the object and the hologram plane [20]. We will assume a point object and the geometrical arrangement shown in Fig. 2.14. We realize from our previous discussions that the hologram of a point consists of a ring pattern of increasing density from the center, which may not lie in the hologram aperture. The number of rings recorded on the

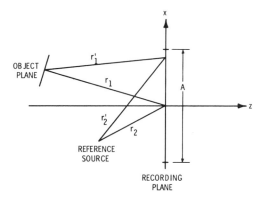

Fig. 2.14. This diagram establishes the geometry in the recording step, for our resolution calculation.

hologram determines the final size of the image point. We prove this statement as follows.

The intensity of the interference pattern in the hologram plane is approximately

$$I_1(x) = |\exp[-ik_1 r_1'] + \exp[-ik_1 r_2']|^2 \qquad (2.71)$$

$$= 2 + \exp[-ik_1(r_1' - r_2')] + \exp[ik_1(r_1' - r_2')] \qquad (2.72)$$

where we have suppressed the time dependence and assumed monochromaticity. Suppose we photograph this interference pattern on an aperture of size A, photographically magnify it to a size mA and then illuminate it with another point source of frequency ω_2 as shown in Fig. 2.15. The amplitude

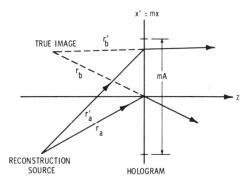

Fig. 2.15. The image reconstruction from a magnified hologram at a different frequency is shown in this drawing.

of the radiation at the true image point becomes

$$s_b(x_b, z_b) = \int_{-\frac{mA}{2}}^{\frac{mA}{2}} \exp\left\{-ik_2\left[r_a' - r_b' + \frac{k_1}{k_2}(r_1' - r_2')_{x'}\right]\right\} dx' \quad (2.73)$$

where $(r_1' - r_2')_{x'}$ denotes that the substitution $x = x'/m$ has been made. We make the usual approximations to the binomial expansions of r_a', r_b', r_1', and r_2' and obtain the expression

$$s_b(x_b, z_b) = \int_{-\frac{mA}{2}}^{\frac{mA}{2}} \exp\Bigg[-ik_2\bigg\{r_a + r_b + \frac{k_1}{k_2}(r_1 - r_2)$$
$$+ \frac{(x')^2}{2}\left[\frac{1}{r_b} + \frac{1}{r_a} + \frac{k_1}{m^2 k_2}\left(\frac{1}{r_1} - \frac{1}{r_2}\right)\right]$$
$$- x'\left[\frac{x_b}{r_b} + \frac{x_a}{r_a} + \frac{k_1}{mk_2}\left(\frac{x_1}{r_1} - \frac{x_2}{r_2}\right)\right]\bigg\}\Bigg] dx' \quad (2.74)$$

Since we wish to find the size of the image point at the image plane, we substitute the image location Eq. (2.54) into Eq. (2.74) and obtain

$$s_b(x_b, z_b) = \int_{-\frac{mA}{2}}^{\frac{mA}{2}} \exp\left\{ik_2x'\left[\frac{x_b}{r_b} + \frac{x_a}{r_a} + \frac{k_1}{mk_2}\left(\frac{x_1}{r_1} - \frac{x_2}{r_2}\right)\right]\right\} dx'$$

$$= mA \frac{\sin\left\{\frac{mk_2A}{2}\left[\frac{x_b}{r_b} + \frac{x_a}{r_a} + \frac{k_1}{mk_2}\left(\frac{x_1}{r_1} - \frac{x_2}{r_2}\right)\right]\right\}}{\frac{mk_2A}{2}\left[\frac{x_b}{r_b} + \frac{x_a}{r_a} + \frac{k_1}{mk_2}\left(\frac{x_1}{r_1} - \frac{x_2}{r_2}\right)\right]} \quad (2.75)$$

The intensity of the image point is $|s_b(x_b, z_b)|^2$, and the zeros of this function occur when

$$\frac{mk_2A}{2}\left[\frac{x_b}{r_b} + \frac{x_a}{r_a} + \frac{k_1}{mk_2}\left(\frac{x_1}{r_1} - \frac{x_2}{r_2}\right)\right] = \pm\pi \quad (2.76)$$

Solving for x_b we obtain

$$x_b = \pm\frac{2\pi r_b}{mk_2A} - r_b\left[\frac{x_a}{r_a} + \frac{k_1}{mk_2}\left(\frac{x_1}{r_1} - \frac{x_2}{r_2}\right)\right] \quad (2.77)$$

The total image point width as measured between zeros of intensity is

$$\delta x_b = \frac{4\pi r_b}{mk_2A} = \lambda_2\frac{r_b}{mA/2} \quad (2.78)$$

The Rayleigh resolution criterion states that two points in object space are resolved if the point of maximum intensity of one image falls on the point of zero intensity of the second image. If we use this directly we find that the resolution in image space is

$$\Delta_b = \frac{\lambda_2 r_b}{mA} \tag{2.79}$$

If we wish to find the object point motion required to move the image by Δ_b, we solve Eq. (2.76) for x_1, with the result that

$$\Delta_1 = \frac{\lambda_1 r_1}{A} \tag{2.80}$$

If we had used a circular aperture, Eq. (2.75) would have come out to be $J_1(\theta)/\theta$ with the result that the resolution expressions are multiplied by the factor 1.22. Normally, holograms are rectangular, mainly because that seems to be the shape in which photographic film is made.

These expressions are not the most desirable in that the resolution is defined in a plane parallel to the x–y plane. This is acceptable if object and image lie near the normal to the hologram. Very often, however, this is not the case so we must proceed to use Eq. (2.76) in a slightly different way. Instead of solving for x_1, we solve for x_1/r_1. That is,

$$\cos \alpha_1 = \frac{x_1}{r_1} = \pm \frac{2\pi}{k_1 A} - \frac{mk_2}{k_1}\left[\frac{x_b}{r_b} + \frac{x_a}{r_a} - \frac{k_1}{mk_2} \right] \tag{2.81}$$

The total variation of $\cos \alpha_1$ required to move the image one Rayleigh resolution distance is

$$\left| \cos \alpha_1' - \cos \alpha_1'' \right| = \frac{2\pi}{k_1 A} = \frac{\lambda_1}{A}$$

where α_1', α_1'' are the limits of the angular motion of the object point. Using a trigonometric identity for the left-hand side, we have

$$\left| \sin\left(\frac{\alpha_1' - \alpha_1''}{2} \right) \right| = \frac{\lambda_1}{2A \left| \sin\left(\dfrac{\alpha_1' + \alpha_1''}{2} \right) \right|}$$

which, for small $\alpha_1' - \alpha_1''$ can be simplified to

$$\Delta\alpha_1 = \left| \alpha_1' - \alpha_1'' \right| = \frac{\lambda_1}{A \sin \alpha_1} \tag{2.82}$$

where α_1 is the average position, $(\alpha_1' + \alpha_1'')/2 < \pi/2$.

This expression shows that the effect of an off-axis object is to reduce the effective aperture. If we now wish to define a resolution in a plane normal to the object vector we can easily do so. That is,

$$\Delta_1' = r_1 \Delta \alpha_1 = \frac{\lambda_1 r_1}{A \sin \alpha_1} \tag{2.83}$$

which reduces to Eq. (2.80) for $\alpha_1 = \pi/2$.

The reader may notice that we were very careful to say that we move the object point by that amount which results in moving the image point by the Rayleigh resolution distance. We never inferred that we had two object points simultaneously. This is because with coherent radiation the phase of the two object points is important. Goodman illustrates this very nicely in his book [19]. Figure 2.16 serves to illustrate the effect of phase on the image. This figure shows that if two points having the same phase and separated by the Rayleigh distance are imaged, they will not be resolved. However, if their phase differs by 180° they will be resolved much better than with incoherent illumination. Only when their phase difference is 90° is the image identical to that under incoherent illumination. To be completely unambiguous, one should probably define a new resolution distance for which the worst case (same phase) is resolved.

The way in which such a phase difference might occur is illustrated in Fig. 2.17. Although the distances from each object point to the origin are equal, the same is not true of their distances from the illumination source. Therefore, the two points have a phase difference of $\Delta \phi = \omega d/c$ which can vary over the whole phase spectrum $0 - 2\pi$. Hence, with this kind of illumination the effects shown in Fig. 2.16 can certainly occur. In normal practice, however, the illumination source is not a perfect point but rather a disc of the order of 10 wavelengths in diameter, or it may be a large diffuse source which can be produced by placing a diffuser in the illuminating beam. In this case, the phase at each point on the object is a summation of

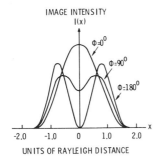

Fig. 2.16. This figure illustrates the pitfalls when we attempt to use the Rayleigh resolution criterion for coherent imaging systems. Two point objects separated by one Rayleigh distance are variously imaged depending on their phase relationship.

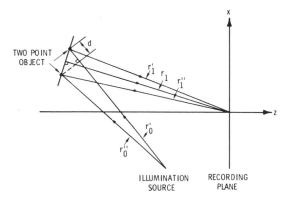

Fig. 2.17. This figure illustrates the way in which two point objects may differ in phase even though they are at equal distance from the origin.

many phases due to rays arriving at the point via many paths. The average phase is then of consequence and can generally be considered zero. Thus, it seems that the Rayleigh criterion is useful after all.

Another useful measure one might like to have is a resolution along the radius to the object; that is, a depth resolution. This may be obtained by substituting the angular object position Eq. (2.55) into Eq. (2.74), leaving the following integral

$$s_b(x_b, z_b) = \int_{-\frac{mA}{2}}^{\frac{mA}{2}} \exp\left\{-ik_2(x')^2\left[\frac{1}{r_b} + \frac{1}{r_a} + \frac{k_1}{m^2k_2}\left(\frac{1}{r_1} - \frac{1}{r_2}\right)\right]\right\} dx'$$

(2.84)

When we make the change of variable,

$$\phi = \left(\frac{2k_2}{\pi}\right)^{\frac{1}{2}}\left[\frac{1}{r_b} + \frac{1}{r_a} + \frac{k_1}{m^2k_2}\left(\frac{1}{r_1} - \frac{1}{r_2}\right)\right]^{\frac{1}{2}} x'$$

we obtain the standard Fresnel integral shown below:

$$\frac{mA}{4Q}\int_0^Q \exp\left[-i\frac{\pi}{2}\phi^2\right] d\phi = \frac{mA}{4}\left[\frac{C(Q) - iS(Q)}{Q}\right]$$

(2.85)

where

$$Q = \frac{mA}{2}\left(\frac{2k_2}{\pi}\right)^{\frac{1}{2}}\left[\frac{1}{r_b} + \frac{1}{r_a} + \frac{k_1}{m^2k_2}\left(\frac{1}{r_1} - \frac{1}{r_2}\right)\right]^{\frac{1}{2}}$$

$$C(Q) = \int_0^Q \cos\left(\frac{\pi}{2}\phi^2\right) d\phi \quad \text{and} \quad S(Q) = \int_0^Q \sin\left(\frac{\pi}{2}\phi^2\right) d\phi$$

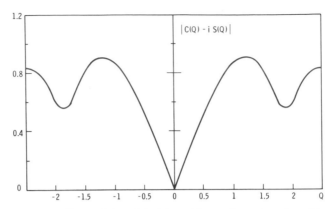

Fig. 2.18. The modulus of the Fresnel integral as a function of its upper limit is shown in this graph.

The Fresnel integral cannot be solved in closed form but it is instructive to plot its value as a function of the limit, Q. Such a plot is shown in Fig. 2.18. By looking at the expression for Q, we see that it becomes zero when the image location Eq. (2.54) is satisfied. At the same time, however, Q also approaches zero in such a way that the ratio approaches unity. Thus, we have the expression

$$s_b(x_b, z_b) \propto \frac{C(Q) - iS(Q)}{Q} \tag{2.86}$$

The image intensity along the radial direction therefore becomes

$$|s_b(x_b, z_b)|^2 \propto \frac{[C(Q)]^2 + [S(Q)]^2}{|Q|^2} \tag{2.87}$$

A plot of this function appears in Fig. 2.19. Note that the plot has no zeros since only nonzero positive quantities are involved. It is customary to allow a 20% loss in intensity from that at the focal plane [20]. From Fig. 2.19 this occurs when $Q \cong \pm 1.0$. Then we have

$$Q = \frac{mA}{2}\left(\frac{2k_2}{\pi}\right)^{\frac{1}{2}}\left[\frac{1}{r_b} + \frac{1}{r_a} + \frac{k_1}{m^2 k_2}\left(\frac{1}{r_1} - \frac{1}{r_2}\right)\right]^{\frac{1}{2}} = \pm 1.0$$

Solving this equation for the spread in r_b for which the intensity remains within the prescribed limits yields

$$\Delta r_b \cong 2\lambda_2\left(\frac{r_b}{mA}\right)^2 \tag{2.88}$$

The amount that the object point could be moved to achieve the same change of intensity of the image at a fixed plane is

$$\Delta r_1 = 2\lambda_1 \left(\frac{r_1}{A} \right)^2 \tag{2.89}$$

A comparison of this expression with Eq. (2.80) shows that the radial resolution is worse than lateral resolution by a factor $2r_1/A$.

We now wish to show that the Rayleigh resolution criterion can be related to the change in the interference ring density on the hologram as the object point is moved. We do this by defining a resolution criterion and then using it to obtain the resolution in object space. A comparison of this expression with the one derived previously shows that they are the same.

We define the resolution Δ_1, in object space as that distance through which the object point may be moved before the total number of interference fringes in the hologram aperture changes by one [''']. This means that the difference in the phases at the aperture extremes changes by 2π. From Eq. (2.72) we know that the phase at any point in the hologram aperture is

$$\phi(x) = k_1(r_1' - r_2') \cong k_1 \left[r_1 - r_2 + \frac{x^2}{2} \left(\frac{1}{r_1} - \frac{1}{r_2} \right) - x \left(\frac{x_1}{r_1} - \frac{x_2}{r_2} \right) \right] \tag{2.90}$$

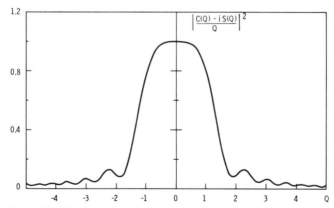

Fig. 2.19. The intensity of the image of a point object along the radius connecting the image to the origin is proportional to the function shown in the graph. A depth of focus can be defined by finding the distance over which image intensity remains above a certain percentage of that at the focal plane. The allowable decrease is usually taken as 20% [20].

The phases at the extremes of the aperture are

$$\phi(+A/2) \cong k_1 \left[r_1 - r_2 + \frac{(A/2)^2}{2} \left(\frac{1}{r_1} - \frac{1}{r_2} \right) - \frac{A}{2} \left(\frac{x_1}{r_1} - \frac{x_2}{r_2} \right) \right] \quad (2.91)$$

and

$$\phi(-A/2) \cong k_1 \left[r_1 - r_2 + \frac{(A/2)^2}{2} \left(\frac{1}{r_1} - \frac{1}{r_2} \right) + \frac{A}{2} \left(\frac{x_1}{r_1} - \frac{x_2}{r_2} \right) \right]$$

The difference between these phases is

$$\delta\phi = k_1 A \left(\frac{x_1}{r_1} - \frac{x_2}{r_2} \right) \quad (2.92)$$

The incremental change of $\delta\phi$ with an incremental change in lateral object position is, for the paraxial approximation,

$$\Delta\delta\phi = \frac{\partial\delta\phi}{\partial x_1} \Delta x_1 = \frac{k_1 A}{r_1} \Delta x_1 \quad (2.93)$$

Setting this expression equal to 2π and solving for Δx_1 results in

$$\Delta_1 = \Delta x_1 = \frac{2\pi r_1}{k_1 A} = \frac{\lambda_1 r_1}{A} \quad (2.94)$$

which is the same as that derived by the Rayleigh criterion. A similar calculation leads to the angular resolution

$$\Delta_\alpha = \frac{\lambda_1}{A \sin \alpha_1} \quad (2.95)$$

and the normal resolution,

$$\Delta_1' = \frac{\lambda_1 r_1}{A \sin \alpha_1} \quad (2.96)$$

These resolution expressions are identical to those based upon the Rayleigh criterion and are somewhat easier to derive. Hence, this definition will be used in this book.

REFERENCES

1. D. Gabor, A new microscope principle, *Nature* **161**:777 (1948).
2. J. B. DeVelis and G. O. Reynolds, *Theory and Applications of Holography*, Addison–Wesley Publishing Co. (1967).

3. E. N. Leith, J. Upatnieks, B. P. Hildebrand, and K. A. Haines, Requirements for a wave front reconstruction television facsimile system, *J. S.M.P.T.E.* **74**(10):893 (1965).

4. B. J. Thompson, A new method of measuring particle size by diffraction techniques, *Japan J. App. Phys.* **4**:302 (1965).

5. A. Lohmann, Optical single-sideband transmission applied to the Gabor microscope, *Opt. Acta* **3**:97 (1956).

6. E. N. Leith and J. Upatnieks, Reconstructed wave fronts and communication theory, *J. Opt. Soc. Am.* **52**:1123 (1962).

7. J. B. DeVelis and G. O. Reynolds, *Theory and Applications of Holography*, Addison–Wesley Publishing Co., Chapt. 2 (1967).

8. M. Born and E. Wolf, *Principles of Optics*, 2nd, Ed., MacMillan Publishing Co., New York, p. 382 (1964).

9. M. Born and E. Wolf, *Principles of Optics*, 2nd Ed., MacMillan Publishing Co., New York, p. 428 (1964).

10. R. F. van Ligten, Influence of photographic film on wavefront reconstruction I; Plane Wavefronts, *J. Opt. Soc. Am.* **56**:1009 (1966).

11. J. R. Varner and J. S. Zelenka, A new method of generating depth contours holographically, *Appl. Opts.* **7**:2107 (1968).

12. M. Born and E. Wolf, *Principles of Optics*, 2nd Ed., MacMillan Publishing Co., New York, p. 491 (1964).

13. M. Born and E. Wolf, *Principles of Optics*, 2nd Ed., MacMillan Publishing Co., New York, p. 504 (1964).

14. E. N. Leith, J. Upatnieks, and K. A. Haines, Microscopy by wavefront reconstruction, *J. Opt. Soc. Am.* **55**(8):981 (1965).

15. E. B. Champagne, Nonparaxial imaging, magnification, and aberration properties in holography, *J. Opt. Soc. Am.* **57**(1):51 (1967).

16. R. W. Meier, Magnification and third-order aberrations in holography, *J. Opt. Soc. Am.* **55**(8):987 (1965).

17. K. A. Haines, The analysis and application of hologram interferometry, Report No. 7421-25-T of the Willow Run Laboratories, Institute of Science and Technology, The University of Michigan, Ann Arbor (1967).

18. M. Born and E. Wolf, *Principles of Optics*, 2nd Ed., MacMillan Publishing Co., New York, p. 166 (1964).

19. J. W. Goodman, *Introduction to Fourier Optics*, McGraw-Hill, New York, Chapter 6 (1968).

20. M. Born and E. Wolf, *Principles of Optics*, 2nd Ed., MacMillan Publishing Co., New York, p. 441 (1964).

21. B. P. Hildebrand and K. A. Haines, Holography by scanning, *J. Opt. Soc. Am.* **59**(1):1 (1969).

Chapter 3

Acoustics

3.1. INTRODUCTION

The term acoustics covers a broad range of subjects, many of which need not be considered in this book. Among those subjects not covered are architectural acoustics, communication acoustics, noise and vibration analysis, music, techniques for measuring elastic constants of solid materials, techniques for studying nonisotropic stresses and inhomogeneities, and measurements of molecular structure of organic liquids.

A very broad spectral range is encompassed within the field of acoustics. Any mechanical vibration with frequencies ranging from a lower limit of 0 Hz to an upper limit of 10^{12} Hz is within the realm of consideration. The reason for the lower limit is obvious, and the upper limit is a consequence of the crystalline atomic structure of solids. The term "acoustic" applies to all mechanical vibrations and compressional and shear waves having any frequency. Similarly, the terms "sound" and "sonic" are not restricted to the audio range.

The subjects discussed in this chapter include a review of mechanical vibrations, mathematical treatment of the propagation of sound in liquids, energy partition at interfaces, radiation pressure, amplitudes of vibration, and interference phenomena.

3.2. MECHANICAL VIBRATIONS

One may obtain a physical conception of sound waves by considering the model of a vibrating system such as that shown in Fig. 3.1. The model consists of a rigidly fixed structure T, a spring S, a massive block B, a dash

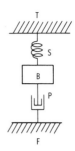

Fig. 3.1. The dynamic behavior of a vibrating system consisting of a
spring, a massive block and a dash pot fixed to rigid-terminal structures
serves as a model for acoustical vibrations.

pot P, and a rigidly fixed base F. Under static conditions the downward
pull of gravity is opposed by the upward pull of the spring. However, if
the block B is displaced downward a distance $y(<0)$ and suddenly released,
the upward force of the spring exceeds the pull of gravity so that the block
accelerates towards T. The equation describing the motion is

$$M \frac{d^2y}{dt^2} = -Ky - R \frac{dy}{dt} \tag{3.1}$$

where M is the mass of the block (kg), g is the acceleration of gravity
(9.8 m/sec^2), K is the spring constant (newton/m), and R is the damping
constant (newton-sec/m). A particular solution for Eq. (3.1) is

$$y = Y \exp(-\alpha t) \cos \omega t \tag{3.2}$$

Upon differentiation we have

$$\frac{dy}{dt} = -Y \exp(-\alpha t)(\omega \sin \omega t + \alpha \cos \omega t) \tag{3.3}$$

and

$$\frac{d^2y}{dt^2} = \alpha Y(\omega \sin \omega t + \alpha \cos \omega t) \exp(-\alpha t) - Y(\omega^2 \cos \omega t - \alpha \omega \sin \omega t) \exp(-\alpha t) \tag{3.4}$$

When Eqs. (3.2), (3.3), and (3.4) are substituted into Eq. (3.1) we have

$$Y\omega(2M\alpha - R) \exp(-\alpha t) \sin \omega t + Y[K - R\alpha + M(\alpha^2 - \omega^2)] \cos \omega t = 0 \tag{3.5}$$

If Eq. (3.5) is to be valid for arbitrary values of time t, we must have

$$\alpha = \frac{R}{2M} \tag{3.6}$$

and

$$M(\omega^2 - \alpha^2) = K - R\alpha \tag{3.7}$$

Equation (3.7) may be rearranged to obtain

$$\omega^2 = \omega_0{}^2 - \alpha^2 \tag{3.8}$$

where

$$\omega_0{}^2 = \frac{K}{M} \tag{3.9}$$

The angular frequency ω_0 is the natural frequency of the vibrating system when no damping forces are present, i.e., when $R = 0$.

It will be useful for future reference to consider the energy of an oscillating system or at least of this system, and similar systems that are simple harmonic in nature. At any given instant, the total energy of the system is partly the potential energy of the spring and partly the kinetic energy of the moving mass. This energy is being dissipated by the dash pot, but at the instant that $y = 0$ the potential energy of the spring vanishes and the kinetic energy of the block represents the total energy of the system. At that time $\cos \omega t = 0$, $\sin \omega t = 1$ and the total energy is given by

$$\text{TE} = \frac{1}{2} M \left(\frac{dy}{dt} \right)^2 = \frac{1}{2} M \omega^2 Y^2 \exp(-2\alpha t_n) \tag{3.10}$$

where t_n is the time at which $y = 0$. An interesting point to note here is the relationship between the angular frequency ω and TE, namely,

$$\omega = \frac{1}{Y} \sqrt{\frac{2\,\text{TE}}{M}} \exp(\alpha t_n) \tag{3.11}$$

3.3. PROPAGATION OF SOUND IN LIQUIDS

3.3.1. The Force Equation

Consider the force on a small volume $\Delta x\, \Delta y\, \Delta z$ of liquid such as shown in Fig. 3.2. Let $P(x, y, z)$ denote the pressure in the liquid at the location (x, y, z). The force F_x on the left-hand side of the cube is

$$F_x = P\, \Delta y\, \Delta z \tag{3.12}$$

while that on the right-hand side is

$$F_x + \frac{\partial F_x}{\partial x}\, \Delta x = \left(P + \frac{\partial P}{\partial x}\, \Delta x \right) \Delta y\, \Delta z \tag{3.13}$$

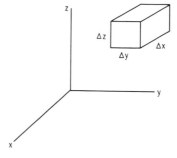

Fig. 3.2. An incremental volume of liquid $\Delta x \Delta y \Delta z$ is used in the derivation of equations describing the propagation of sound in liquids.

Thus, the net force in the x direction is

$$\Delta F_x = -\frac{\partial F_x}{\partial x}\Delta x = -\frac{\partial P}{\partial x}\Delta x\,\Delta y\,\Delta z \qquad (3.14)$$

Similarly,

$$\Delta F_y = -\frac{\partial P}{\partial y}\Delta x\,\Delta y\,\Delta z \qquad (3.15)$$

and

$$\Delta F_z = -\frac{\partial P}{\partial z}\Delta x\,\Delta y\,\Delta z \qquad (3.16)$$

Taking into consideration all of these forces, we find that the net force per unit volume is given by

$$\mathbf{F} = (\mathbf{i}\Delta F_x + \mathbf{j}\Delta F_y + \mathbf{k}\Delta F_z)/\Delta x\,\Delta y\,\Delta z \qquad (3.17)$$

where $\mathbf{i}$, $\mathbf{j}$, $\mathbf{k}$ are unit vectors in the x, y, and z directions, respectively. It follows that

$$\mathbf{F} = -\nabla P \qquad (3.18)$$

Although the preceding equations were derived on the basis of P representing the total pressure in the liquid at any given point, the equations remain valid when, instead of P, we use p, the difference between the total dynamic pressure and the pressure P_0 at the same point under static conditions:

$$P = P_0 + \Delta P = P_0 + p \qquad (3.19)$$

Since P_0 is a constant

$$\nabla P = \nabla p \qquad (3.20)$$

so

$$\mathbf{F} = -\nabla p \qquad (3.21)$$

3.3.2. Description of the Wave

A sound wave of plane wave front moving in the z direction may be described by

$$w = W \exp i(\Omega t - Kz) \tag{3.22}$$

where it is understood that only the real part of the complex expression is used. The wave described by Eq. (3.22) is traveling in the positive z direction, has an angular frequency Ω and a wave number

$$K = \frac{2\pi}{\Lambda} \tag{3.23}$$

where Λ is the wavelength. The particle displacement in the liquid at a given instant t and position z is denoted by w. The displacement amplitude is denoted by W.

3.3.3. Compressibility

One of the characteristics of a liquid which relates to the propagation of sound is its compressibility which we shall denote by the symbol β. By definition, β_i is the change in volume, ∂V, per unit volume, V, produced by a change in pressure, ∂P, while the whole volume is maintained at constant temperature:

$$\beta_i = -\frac{1}{V} \left(\frac{\partial V}{\partial P} \right)_T \tag{3.24}$$

The subscript i is used to indicate that this is the isothermal compressibility [1]. In sound propagation, adiabatic conditions are more likely to prevail. The adiabatic compressibility β_a is linearly related to the isothermal compressiblity and we may write

$$\beta_a = \beta_i / \varkappa \tag{3.25}$$

where $\varkappa^{-1}$ is the constant of proportionality. Thus we may also write

$$\beta_a = -\frac{1}{\varkappa V} \left(\frac{\partial V}{\partial P} \right)_T \tag{3.26}$$

3.3.4. The Wave Description in Terms of Pressure

Consider an elementary volume $\Delta x \, \Delta y \, \Delta z$ in an undisturbed liquid. This volume is disturbed as shown in Fig. 3.3 when the wave described by Eq. (3.22) passes through it. The bottom surface is displaced a distance w

Fig. 3.3. An element of liquid is either expanded or compressed as a sound wave passes through it. In the case under consideration, where the sound wave is traveling in the z-direction it is the expansion or contraction of the z-dimension that produces the change in volume.

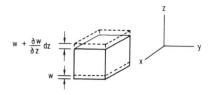

in the z direction, while the top surface is displaced a slightly different distance $w + (\partial w/\partial z)\, \Delta z$. The change in volume ΔV which occurred is

$$\Delta V = \Delta x\, \Delta y\, \Delta z \left(1 + \frac{\partial w}{\partial z}\right) - \Delta x\, \Delta y\, \Delta z$$

$$= \Delta x\, \Delta y\, \Delta z \frac{\partial w}{\partial z} = V \frac{\partial w}{\partial z} \tag{3.27}$$

Thus we can rewrite Eq. (3.26) as

$$p = -\frac{1}{\beta_a} \frac{\partial w}{\partial z} \tag{3.28}$$

where, as in Eq. (3.19), $p = \Delta P$. Thus using Eq. (3.22)

$$p = \frac{iK}{\beta_a} W \exp i(\Omega t - Kz) \tag{3.29}$$

that is,

$$p = p_0 \sin(\Omega t - Kz) \tag{3.30}$$

where the pressure amplitude p_0 is given by

$$p_0 = -\frac{KW}{\beta_a} \tag{3.31}$$

3.3.5. Velocity of Propagation

The volume $\Delta x\, \Delta y\, \Delta z$ of Figs. 3.2 and 3.3 experiences an acceleration $\partial^2 w/\partial t^2$ because of the force per unit volume given by $-\partial p/\partial z$. Thus we may write

$$-\frac{\partial p}{\partial z} = \varrho \frac{\partial^2 w}{\partial t^2} \tag{3.32}$$

because of Newton's second law. From Eq. (3.30) we have

$$-\frac{\partial p}{\partial z} = p_0 K \cos(\Omega t - Kz) \tag{3.33}$$

and from Eq. (3.22) (real part)

$$\frac{\partial w}{\partial t} = -\Omega W \sin(\Omega t - Kz) \tag{3.34}$$

and

$$\frac{\partial^2 w}{\partial t^2} = -\Omega^2 W \cos(\Omega t - Kz) \tag{3.35}$$

From these equations we readily see that

$$p_0 K = - \varrho \Omega^2 W \tag{3.36}$$

or

$$\frac{K^2}{\beta} W = \varrho \Omega^2 W \tag{3.37}$$

or

$$\frac{\Omega^2}{K^2} = \frac{1}{\varrho \beta} \tag{3.38}$$

But

$$\frac{\Omega^2}{K^2} = c^2$$

where c is the velocity of the wave, so

$$c = \frac{1}{\sqrt{\varrho \beta}} \tag{3.39}$$

3.3.6. Radiation Pressure

Propagation of an acoustic wave in a liquid involves some second order effects which produce a small but constant pressure Π in the direction of propagation of the wave [2]. When a small element of volume V is diminished by an amount ΔV in the presence of a sound wave, the density is increased by an amount

$$\Delta \varrho = -\varrho \frac{\Delta V}{V} \tag{3.40}$$

Modifying Eq. (3.24) to

$$\beta = - \frac{\Delta V}{V} \left(\frac{1}{p} \right) \tag{3.41}$$

we also see that

$$\beta = + \frac{\Delta \varrho}{\varrho} \left(\frac{1}{p} \right) \tag{3.42}$$

Also

$$p = \frac{-KW}{\beta} \sin(\Omega t - Kz) \tag{3.43}$$

or

$$p = -\frac{K}{\Omega} \frac{W}{\beta} \Omega \sin(\Omega t - Kz) \tag{3.44}$$

But

$$\frac{K}{\Omega} = \frac{1}{c} \tag{3.45}$$

$$\beta = \frac{1}{\varrho c^2} \tag{3.46}$$

and

$$-W\Omega \sin(\Omega t - Kz) = \frac{dw}{dt} \tag{3.47}$$

so

$$p = \varrho c \frac{dw}{dt} = \varrho c U \tag{3.48}$$

if we denote dw/dt by U. The second-order effect that produces radiation pressure is accounted for if we do not accept ϱ as a constant as is normally done, but rather account for variations in ϱ by writing

$$\varrho = \varrho_0 + \Delta \varrho = \varrho_0 (1 + p\beta) \tag{3.49}$$

We may use the approximate value $\varrho_0 c U$ for p in Eq. (3.49) and determine a more exact value from

$$p = \varrho_0 (1 + \varrho_0 c \beta U) c U \tag{3.50}$$

or using Eq. (3.47) for U

$$p = -\varrho_0 c \Omega W \sin(\Omega t - Kz) + \varrho_0{}^2 c^2 \beta \Omega^2 W^2 \sin^2(\Omega t - Kz) \tag{3.51}$$

The magnitude of the radiation pressure may be taken as the average value of p over one period. If the pressure p were truly sinusoidal, its time average would be zero, but because the density ϱ is not truly a constant there is a steady pressure component given by

$$\Pi = \frac{1}{T} \int_0^T p \, dt \tag{3.52}$$

where

$$T = \frac{2\pi}{\Omega} \tag{3.53}$$

When the integral of Eq. (3.52) is evaluated we obtain

$$\Pi = \tfrac{1}{2}\varrho_0{}^2 c^2 \beta \Omega^2 W^2 \tag{3.54}$$

which, because of Eq. (3.39) may also be written

$$\Pi = \tfrac{1}{2}\varrho_0 \Omega^2 W^2 \tag{3.55}$$

Equation (3.36) may also be written

$$p_0 = -\varrho_0 c \Omega W \tag{3.56}$$

so that

$$\Pi = \frac{p_0{}^2}{2\varrho_0 c^2} \tag{3.57}$$

A perfect absorber in the acoustic beam has a force exerted upon it equal to the product of the radiation pressure and the area projected normal to the beam. A perfect reflector has twice the force per unit area exerted on it because whereas an absorber reduces the momentum of the wave to zero, a reflector produces a wave of equal momentum in the opposite direction. Thus, the reflector produces a change in momentum twice as great as the absorber and since force is proportional to the rate of change of momentum (Newton's second law) the reflector has twice the force per unit area exerted on it than has the absorber.

3.3.7. Energy and Intensity

An infinitesimal volume of liquid $\Delta x\,\Delta y\,\Delta z$ in the sound field has, at any time t, a velocity dw/dt given by Eq. (3.34). Thus the volume has a kinetic energy per unit volume given by

$$KE = \tfrac{1}{2}\varrho \Omega^2 W^2 \sin^2(\Omega t - Kz) \tag{3.58}$$

It also has a potential energy, and the sum of the kinetic and potential energy is the total energy. The total energy per unit volume remains constant if we assume no losses due to absorption. When the potential energy is maximum the kinetic energy is zero and vice versa. Thus the total energy per unit volume of the wave may be taken to be

$$E = \tfrac{1}{2}\varrho \Omega^2 W^2 \tag{3.59}$$

This energy flows through a unit area normal to the direction of propagation of the wave with a velocity c. We define acoustic intensity, I, as the rate of flow of energy through a unit area. Hence

$$I = \tfrac{1}{2}\varrho c \Omega^2 W^2 \tag{3.60}$$

3.3.8. Velocity Potential

Equation (3.32) could well have been written

$$-\text{grad } p = \varrho \, \frac{\partial^2 w}{\partial t^2} \tag{3.61}$$

and using Eq. (3.22) we could also write

$$-\text{grad } p = i\Omega\varrho \, \frac{\partial w}{\partial t} \tag{3.62}$$

If we define a quantity ψ such that

$$p = -i\Omega\varrho\psi \tag{3.63}$$

we find, on substitution into Eq. (3.62), that

$$\text{grad } \psi = \frac{\partial w}{\partial t} \tag{3.64}$$

The quantity ψ is commonly known as the velocity potential.

3.4. PROPAGATION OF SOUND IN SOLIDS

The understanding and analysis of the propagation of sound in solids is complicated by the fact that two types of waves and combinations of these types can be propagated. The two types are compressional or longitudinal, and shear waves. If the solid in which these waves are propagating has a suitable geometry, the two types of waves combined may generate Lamb waves [3].

The proper description of the propagation of sound in solids requires an analytical treatment far beyond that which is reasonable to include in this book. The subject is treated thoroughly by several authors [4,5]. Here we simply take note of the fact that two types of wave, namely longitudinal and shear, can propagate in a solid.

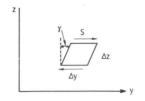

Fig. 3.4. The ratio of the shear stress S to the shear strain γ is known as the modulus of rigidity μ; i.e., $\mu = S/\gamma$.

A longitudinal wave propagates in solids in much the same manner as in liquids. Its velocity, v_l, may be expressed in terms of the Lamé constants, λ and μ, as follows

$$v_l = \sqrt{\frac{\lambda + 2\mu}{\varrho}} \tag{3.65}$$

where ϱ is the density of the solid [5]. The second Lamé constant μ is readily defined by referring to Fig. 3.4 in which an elemental volume is subjected to a shear stress S. The shear stress is the tangential stress per unit area. A modulus of rigidity or shear modulus μ is defined as the ratio of shear stress to the shear strain:

$$\mu = \frac{S}{\gamma} \tag{3.66}$$

where γ is the angular displacement of the ends of the elemental volume due to the shearing stress S.

As a means of comparison between liquids and solids it will be useful to relate the bulk modulus, or modulus of compressibility, of a solid to the Lamé constants. This is done through the equation

$$\beta^{-1} = \frac{3\lambda + 2\mu}{3} \tag{3.67}$$

In liquids, where shear rigidity does not exist, the bulk modulus reduces to

$$\beta^{-1} = \lambda \tag{3.68}$$

In solids the shear modulus affects the bulk modulus, and the velocity of longitudinal waves cannot be expressed in terms of the density and bulk modulus alone as it can for a liquid.

Shear waves in solids propagate with a velocity, v_s, given by (5)

$$v_s = \sqrt{\frac{\mu}{\varrho}} \tag{3.69}$$

3.4.1. Refraction and Reflection at Liquid–Solid Interfaces

If a plane acoustic wave is incident from the liquid side of a liquid–solid interface at an angle $\alpha(IL)$ to the normal, some of the energy will be reflected at an opposite but equal angle $\alpha(RL)$, some will be transmitted through the interface as a longitudinal wave in the solid and be refracted at an angle $\alpha(TL)$ and some will be transmitted as a shear wave and be refracted at an angle $\alpha(TS)$ [7,8]. The situation is diagramed in Fig. 3.5. Figure 3.5 is drawn for the most common case in which the relative magnitudes of the velocities are $v_l > v_s > c$. The well-known laws from optics (i.e., Snells law and the law of reflection) dictate that these velocities and angles of propagation be related by the equations

$$\frac{c}{\sin \alpha(IL)} = \frac{v_l}{\sin \alpha(TL)} = \frac{v_s}{\sin \alpha(TS)} = \frac{-c}{\sin \alpha(RL)} \tag{3.70}$$

Let the velocity potential amplitudes of the waves shown in Fig. 3.5 be IL, RL, RS, TL, and TS. The manner in which the incident energy is partitioned is given by the following equations [7]:

$$\frac{RL}{IL} = \frac{Z(L2) \cos^2 2\alpha(TS) + Z(S2) \sin^2 2\alpha(TS) - Z(L1)}{Z(L2) \cos^2 2\alpha(TS) + Z(S2) \sin^2 2\alpha(TS) + Z(L1)} \tag{3.71}$$

$$\frac{RS}{IL} = 0 \tag{3.72}$$

$$\frac{TL}{IL} = \frac{\varrho_1}{\varrho_2} \frac{2Z(L2) \cos 2\alpha(TS)}{Z(L2) \cos^2 2\alpha(TS) + Z(S2) \sin^2 2\alpha(TS) + Z(L1)} \tag{3.73}$$

$$\frac{TS}{IL} = -\frac{\varrho_1}{\varrho_2} \frac{2Z(S2) \sin 2\alpha(TS)}{Z(L2) \cos^2 2\alpha(TS) + Z(S2) \sin^2 2\alpha(TS) + Z(L1)} \tag{3.74}$$

where the impedances Z are given by

$$Z(L2) = \frac{\varrho_2 v_l}{\cos \alpha(TL)} \tag{3.75}$$

$$Z(S2) = \frac{\varrho_2 v_s}{\cos \alpha(TS)} \tag{3.76}$$

$$Z(L1) = \frac{\varrho_1 c}{\cos \alpha(IL)} \tag{3.77}$$

where ϱ_1 is the density of the liquid and ϱ_2 is the density of the solid.

At near normal incidence where $\alpha(IL)$ and hence $\alpha(TS)$ are small, we see that the transmitted shear wave amplitude (TS) is much less than the

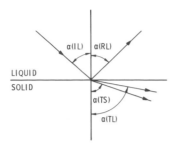

Fig. 3.5. An acoustical wave incident from the liquid side of a liquid–solid interface partitions its energy among three waves when it interacts with the liquid–solid interface. Of these three, two are formed by refraction and one by reflection.

transmitted longitudinal wave amplitude. At true normal incidence

$$\alpha(IL) = \alpha(TL) = \alpha(TS) = 0 \tag{3.78}$$

$$\frac{RL}{IL} = \frac{Z(L2) - Z(L1)}{Z(L2) + Z(L1)} \tag{3.79}$$

$$\frac{TL}{IL} = \frac{\varrho_1}{\varrho_2} \frac{2Z(L2)}{Z(L2) + Z(L1)} \tag{3.80}$$

$$\frac{TS}{IL} = 0 \tag{3.81}$$

When a longitudinal wave is incident upon the solid–liquid interface from the solid side there will be a transmitted longitudinal, a reflected longitudinal, and a reflected shear wave as shown in Fig. 3.6 for the case in which the relative magnitudes of the velocities are as before, $v_l < v_s < c$. The incident energy is partitioned in a manner which can be calculated from

$$\frac{RL}{IL} = \frac{Z(L1) + Z(S2)\sin^2 2\alpha(RS) - Z(L2)\cos^2 2\alpha(RS)}{Z(L1) + Z(S2)\sin^2 2\alpha(RS) + Z(L2)\cos^2 2\alpha(RS)} \tag{3.82}$$

$$\frac{TL}{IL} = \left(\frac{\cos \alpha(IL)}{\cos \alpha(TL)\cos^2 2\alpha(RS)}\right)\left(1 - \frac{RL}{IL}\right)\frac{c}{v_l} \tag{3.83}$$

$$\frac{RS}{IL} = \left(\frac{\sin 2\alpha(IL)}{\cos 2\alpha(RS)}\right)\left(1 - \frac{RL}{IL}\right)\left(\frac{v_s}{v_l}\right)^2 \tag{3.84}$$

Fig. 3.6. An acoustical wave incident from the solid side of a liquid–solid interface partitions its energy among three waves when it interacts with the liquid–solid interface. Of these three, two are formed by reflection and one by refraction.

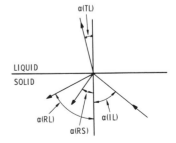

At normal incidence

$$\frac{RL}{IL} = \frac{Z(L1) - Z(L2)}{Z(L2) + Z(L1)} \tag{3.85}$$

and

$$\frac{TL}{IL} = \frac{2Z(L2)}{Z(L2) + Z(L1)} \frac{c}{v_l} \tag{3.86}$$

These equations have the same form as Eqs. (3.79) and (3.80), but the roles of the impedances Z have been interchanged so that

$$Z(L1) = \frac{\varrho_1 c}{\cos \alpha(TL)} \tag{3.87}$$

and

$$Z(L2) = \frac{\varrho_2 v_l}{\cos \alpha(IL)} \tag{3.88}$$

Finally, there is the case of a transverse (shear) wave incident from a solid onto a solid–liquid interface. If the particle velocity is parallel to the boundary of separation, no longitudinal waves will be generated in either the solid or the liquid upon reflection. This component of the shear wave is totally reflected.

If the particle velocity lies in the plane of incidence the manner in which the energy is partitioned may be calculated from

$$\frac{RS}{IS} = - \frac{Z(L1) + Z(L2) \cos^2 2\alpha(IS) - Z(S2) \sin^2 2\alpha(IS)}{Z(L1) + Z(L2) \cos^2 2\alpha(IS) + Z(S2) \sin^2 2\alpha(IS)} \tag{3.89}$$

$$\frac{TL}{IS} = \frac{\tan \alpha(TL)}{2 \sin^2 \alpha(IS)} \left(1 + \frac{RS}{IS}\right) \tag{3.90}$$

$$\frac{RL}{IS} = - \frac{\cos 2\alpha(IS)}{\sin 2\alpha(RL)} \left(1 + \frac{RS}{IS}\right) \left(\frac{v_l}{v_s}\right)^2 \tag{3.91}$$

At normal incidence

$$\frac{RS}{IS} = -1 \tag{3.92}$$

$$\frac{TL}{IS} = \frac{RL}{IS} = 0 \tag{3.93}$$

In all the cases that we have discussed we assumed that the waves described are the only ones which are generated, so all the energy resides in these waves. For the case in which the incident wave is longitudinal and incident from the liquid side, conservation of energy requires that

$$\frac{\varrho_1{}^2}{Z(L1)}\,(RL)^2 + \frac{\varrho_2{}^2}{Z(L2)}\,(TL)^2 + \frac{\varrho_2{}^2}{Z(S2)}\,(TS)^2 = \frac{\varrho_1{}^2}{Z(L1)}\,(IL)^2 \quad (3.94)$$

Similarly, when the incident wave is longitudinal and incident from the solid

$$\frac{\varrho_2{}^2}{Z(L2)}\,(RL)^2 + \frac{\varrho_2{}^2}{Z(S2)}\,(RS)^2 + \frac{\varrho_1{}^2}{Z(L1)}\,(TL)^2 = \frac{\varrho_2{}^2}{Z(L2)}\,(IL)^2 \quad (3.95)$$

When the incident wave is transverse with oscillations in the plane of incidence and incident from the solid

$$\frac{\varrho_2{}^2}{Z(S2)}\,(RS)^2 + \frac{\varrho_1{}^2}{Z(L1)}\,(TL)^2 + \frac{\varrho_2{}^2}{Z(L2)}\,(RL)^2 = \frac{\varrho_2{}^2}{Z(S2)}\,(IS)^2 \quad (3.96)$$

3.5. INTERACTION OF AN ACOUSTIC WAVE WITH A LIQUID INTERFACE

3.5.1. Interaction at a Free-Liquid Surface

In Section 3.3.6 we saw that when a beam of acoustic energy impinges upon a perfect reflector at normal incidence, a pressure given by

$$\Pi_r = \frac{p_0{}^2}{\varrho_0 c^2} \quad (3.97)$$

is exerted upon the surface. In terms of the acoustic intensity, which by comparison of Eqs. (3.56) and (3.60) may be written

$$I = \frac{p_0{}^2}{2\varrho c} \quad (3.98)$$

this pressure is given by

$$\Pi_r = \frac{2I}{c} \quad (3.99)$$

The effect of the pressure upon the surface of the liquid is to distort it an amount $z' = f(x, y)$ where the prime is used to distinguish the function $f(x, y)$ from the coordinate z. Consider the forces acting on a small volume, dv, at a position (x, y, z'). We assume that the force is due to a continuous wave with an amplitude that varies with x and y but not with time. The volume element will be in equilibrium under the action of forces due to gravitation, the surface free energy (surface tension) and the acoustical radiation pressure. In the case of a plane wave of constant amplitude and large area, the surface tension effect is negligible and the radiation pressure Π is balanced by the gravitational pressure $\varrho gz'$, so that

$$\Pi = \varrho gz' \tag{3.100}$$

Using Eq. (3.99) and solving for z' we find that

$$z' = \frac{2I}{\varrho gc} \tag{3.101}$$

A case in which the surface tension is not negligible will be treated in Chapter 6.

It will be useful to compare the magnitude of displacement z' with the particle displacement amplitude w of the wave producing it. Assume a wave of intensity $I = 0.1$ W/cm² and frequency $v = 10$ MHz. In water for which $c = 1.5 \times 10^5$ cm/sec and $\varrho = 1$ g/cm³,

$$z' = 1.36 \times 10^{-2} \text{ cm}$$

Setting Π_r of Eq. (3.99) equal to twice the value of Π in Eq. (3.55) we see that

$$\varrho \Omega^2 W^2 = \frac{2I}{c} \tag{3.102}$$

or

$$W = \frac{1}{2\pi v} \sqrt{\frac{2I}{\varrho c}} \tag{3.103}$$

Using the same values of I and c as in calculating the displacement z', and assuming a frequency $v = 10$ MHz the resulting amplitude W is

$$W = 5.8 \times 10^{-8} \text{ cm}$$

Thus the displacement due to radiation pressure is far greater than that associated with the particle displacement at ultrasonic frequencies.

3.6. SUMMARY

The purpose of this chapter was to present the fundamental concepts relating to acoustics in a concise manner. Equations derived and presented here define the total energy, intensity, pressure and displacement amplitudes, velocity, radiation pressure, and velocity potential of a wave. In addition to these definitions and the equations describing the relationships between these quantities, we included a discussion of energy partition at a liquid–solid interface.

REFERENCES

1. T. F. Hueter and R. H. Bolt, *Sonics*, John Wiley and Sons, Inc., New York, p. 433 (1955).
2. T. F. Hueter and R. H. Bolt, *Sonics*, John Wiley and Sons, Inc., New York, pp. 43–44 (1955).
3. L. A. Viktorov, *Rayleigh and Lamb Waves*, Plenum Press, New York (1967).
4. P. M. Morse, *Vibration and Sound*, McGraw Hill, New York (1948).
5. E. U. Condon and H. Odishaw (eds.), *Handbook of Physics*, McGraw Hill, New York, Part 3, Chapter 7 (1958).
6. L. M. Brekhovskikh, *Waves in Layered Media*, Academic Press, New York, Chapter 1, Sect. 4 and p. 28 (1960).
7. L. M. Brekhovskikh, *Waves in Layered Media*, Academic Press, New York, Chapter 1, Section 4 (1960).
8. W. G. Mayer, Energy partition of ultrasonic waves at flat boundaries, *Ultrasonics*, April–June, pp. 62–68 (1965).

Scanned Acoustical Holography

The preceding chapters have provided an adequate background in both holography and acoustics. We can now turn to the specific application of holography to the visualization of acoustical fields. We restrict our discussion for the most part to longitudinal waves, although our results are applicable to shear waves. We stress that one or the other be used, although there are certain circumstances under which both may be present. We consider this to be unfortunate in that interfering images may result.

4.1. SCANNED RECEIVER

In our earlier discussions, we stated that at long wavelengths we can use a point receiver to scan the field, measure the phase and amplitude at each position, and use this signal to modulate a synchronously scanned light source over unexposed film. The developed film then constitutes a hologram.

4.1.1. Acoustical Reference

The manner in which the phase is measured is a matter of choice. For the researcher familiar with optical holography, the most direct method is to use two acoustical beams. Unlike optics, these two beams can be derived from separate transducers driven by the same oscillator. The beams are arranged so that they overlap in some region of space. The point receiver is then scanned over a surface within the overlap region, the received signal is square-law detected and recorded. For the moment we assume that the recording is electronic, magnetic tape for instance, and is later printed out or written in a format suitable for reconstruction. Figure 4.1 is a schematic of this system. When we refer to the receiver as being a point, we mean that it is smaller than the smallest fringe spacing existing in the interference pattern. Since the fringe spacing can be expressed as

$$d = \frac{\lambda}{2\sin(\theta/2)\sin\phi}$$

where θ is the angle between the two propagation vectors and ϕ is the angle between the plane of the hologram and the bisector of θ, we see that the minimum spacing is $\lambda/2$ when $\theta = \pi$ and $\phi = \pi/2$. Therefore, as long as the receiver is smaller than $\lambda/2$, we can expect to detect all interference fringes that can possibly be generated.

The acoustical reference method results in a signal exactly equivalent to that found on an optical hologram. For completeness we repeat the results of Chapter 2 adapted to the format in this chapter. The amplitude of the acoustic signal in the scanned plane is

$$s(x, y) = s_1(x, y) + s_2(x, y) \tag{4.1}$$

where

$$s_1(x, y) = a_1(x, y)\cos[\omega_1 t + \phi_1(x, y)]$$

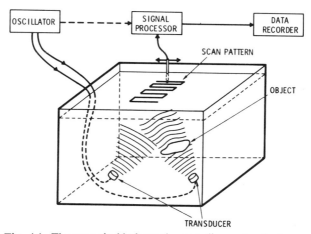

Fig. 4.1. The acoustical holography recording system is shown in schematic form. The signal processor performs the amplitude and phase measurements. It may consist of a square-law detector and low-pass filter when an acoustical reference is used, and a phase detector if none is present.

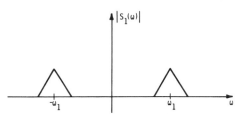

Fig. 4.2. This plot shows the spectrum of the
signal at the detector.

and

$$s_2(x, y) = a_2(x, y) \cos[\omega_1 t + \phi_2(x, y)]$$

are the complex amplitudes of the object and reference beams, respectively. This is the signal that will be produced by the receiver when it is stationary. When the receiver is in motion, both the amplitude and phase terms also become functions of time and if the motion is fast enough, serious problems result. As an example consider only one term of Eq. (4.1) and suppose that we have linear motion of velocity, v, in the x-direction. Then the received signal is

$$s_1(t) = a_1(vt, y) \cos[\omega_1 t + \phi_1(vt, y)]$$

$$= \frac{a_1(vt, y)}{2} \{\exp i[\omega_1 t + \phi_1(vt, y)] + \exp -i[\omega_1 t + \phi_1(vt, y)]\} \quad (4.2)$$

It is often informative to take the Fourier transform of a time varying function. When we do so we obtain

$$S_1(\omega) = F_\omega[s_1(t)] = \int_{-\infty}^{\infty} s_1(t) \exp(-i\omega t)\, dt$$

$$= \int \frac{a_1(vt, y)}{2} \exp[i\phi_1(vt, y)] \exp[-i(\omega - \omega_1)t]\, dt$$

$$+ \int \frac{a_1(vt, y)}{2} \exp[-i\phi_1(vt, y)] \exp[-i(\omega + \omega_1)t]\, dt$$

$$= \tfrac{1}{2} F_{\omega-\omega_1}\{a_1(vt, y) \exp[i\phi_1(vt, y)]\}$$

$$+ \tfrac{1}{2} F_{\omega+\omega_1}\{a_1(vt, y) \exp[i\phi_1(vt, y)]\}^* \quad (4.3)$$

This equation shows that the electronic signal has a spectrum peaked about $\omega = \pm \omega_1$ as shown in Fig. 4.2. The spectral distribution about $\pm \omega_1$ represents the information of the phase and amplitude distribution in the aperture. We get some idea of the effect of scan velocity by expanding the phase term in a Maclaurin series as shown below:

$$\phi_1(vt, y) = \phi_1(0, y) + vt\dot{\phi}_1(0, y) + (vt)^2 \frac{\ddot{\phi}_1(0, y)}{2!} + \cdots \qquad (4.4)$$

where $\dot{\phi}_1(0, y) = \partial\phi(vt, y)/\partial vt$ evaluated at $vt = 0$. If we assume that second and higher derivatives are small compared with the first, we obtain

$$F_{\omega-\omega_1}\{a_1(vt, y) \exp[i\phi_1(vt, y)]\}$$
$$= \exp[i\phi_1(0, y)] \int a_1(vt, y) \exp\{-i[\omega - \omega_1 - v\dot{\phi}_1(0, y)]t\} \, dt$$
$$= \exp[i\phi_1(0, y)]F_{\omega-\omega_1-v\dot{\phi}_1(0,y)} [a(vt, y)] \qquad (4.5)$$

This expression shows that the signal spectrum is the amplitude spectrum shifted to $\omega = \omega_1 + v\dot{\phi}_1(0, y)$. The term $\dot{\phi}_1(0, y)$ represents the spatial frequency with respect to the variable, x, of the phase variation in the aperture, and this, in turn, depends upon the temporal frequency of the source and the angle of inclination of the beam to the aperture. We know that it is possible for the phase pattern to extend from 0 to $\lambda_1/2$ in period; thus $\dot{\phi}_1$ may vary from 0 to $\pm \omega_1(v/c)$. This means that each frequency of the amplitude spectrum will be shifted by a different amount, resulting in a distorted image, as discussed in Section 4.4.3. If the amplitude spectrum is not of great importance, as is usually the case, this distortion will be acceptable.

The total signal as produced by the scanning detector is

$$s(vt, y) = s_1(vt, y) + s_2(vt, y) \qquad (4.6)$$

The output of the square-law detector becomes

$$y(t) = s^2(vt, y) = a_1{}^2(vt, y) \cos^2[\omega_1 t + \phi_1(vt, y)]$$
$$+ a_2{}^2(vt, y) \cos^2[\omega_1 t + \phi_2(vt, y)]$$
$$+ 2a_1(vt, y)a_2(vt, y) \cos[\omega_1 t + \phi_1(vt, y)] \cos[\omega_1 t + \phi_2(vt, y)] \qquad (4.7)$$

Expanding the cosine terms by means of the appropriate trigonometric identities we obtain

$$y(t) = \tfrac{1}{2}\{[a_1{}^2(vt, y) + a_2{}^2(vt, y)] + a_1{}^2(vt, y) \cos[2\omega_1 + 2\phi_1(vt, y)]$$
$$+ a_2{}^2(vt, y) \cos[2\omega_1 t + 2\phi_2(vt, y)]$$
$$+ 2a_1(vt, y)a_2(vt, y)(\cos[2\omega_1 + \phi_1(vt, y) + \phi_2(vt, y)]$$
$$+ \cos[\phi_1(vt, y) - \phi_2(vt, y)])\} \qquad (4.8)$$

We note that with the exception of the first and last terms the expression represents a signal modulated onto a carrier of frequency $2\omega_1$. A square-law device used in this manner is always followed by a low-pass filter which has the effect of rejecting all those terms on the carrier. The signal for one scan line that is finally recorded is therefore

$$z(t) = \tfrac{1}{2}\{a_1{}^2(vt, y) + a_2{}^2(vt, y)$$
$$+ 2a_1(vt, y)a_2(vt, y) \cos[\phi_1(vt, y) - \phi_2(vt, y)]\} \qquad (4.9)$$

The total signal is the sum of N similar signals at N values of y. That is, the receiver is stepped in the y-dimension after each scan in the x-direction is completed.

4.1.2. Electronic Reference

As we mentioned earlier, it is not necessary to provide an acoustical reference when using the scanning method. In fact, there is a distinct advantage in not doing so, which we shall elaborate on later. The alternative to the acoustical reference is an electronically simulated one. For this type of system the signal sent out by the scanning receiver is

$$s(vt, y) = a_1(vt, y) \cos[\omega_1 t + \phi_1(vt, y)] \qquad (4.10)$$

This signal is fed to a "black box" designated as the signal processor. The major function of the signal processor is to detect the amplitude and phase of the signal. This can be done in several ways, some of which are outlined below.

One method of phase detection is shown in schematic form in Fig. 4.3. The input signal is multiplied by a portion of the signal used to drive the transducer. This results in a signal

$$y(t) = a_0 a_1(vt, y) \cos[\omega_1 t + \phi_1(vt, y)] \cos \omega_1 t$$
$$= \frac{a_0 a_1(vt, y)}{2} \{\cos[2\omega_1 t + \phi_1(vt, y)] + \cos[\phi_1(vt, y)]\} \qquad (4.11)$$

Upon passing through the low-pass filter, the final signal is

$$z(t) = \frac{a_0 a_1(vt, y)}{2} \cos \phi_1(vt, y) \qquad (4.12)$$

This expression shows that we have indeed succeeded in preserving all information in the object beam; and in a "clean" form. In contrast, the

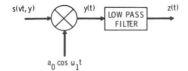

Fig. 4.3. One form of phase detection is shown in this schematic. This is actually the preferred method since it results in a hologram without a space-varying interference term.

acoustic reference method resulted in preservation of all information but "dirtied" by other factors as in Eq. (4.9). The significance of this was alluded to earlier and will be elaborated later.

A second popular detection method is shown in Fig. 4.4. For this case, $y(t)$ becomes

$$y(t) = a_0 \cos \omega_1 t + a_1(vt, y) \cos[\omega_1 t + \phi_1(vt, y)] \qquad (4.13)$$

The output of the square-law detector is

$$x(t) = \tfrac{1}{2}\{[a_0{}^2 + a_1{}^2(vt, y)] + a_0{}^2 \cos 2\omega_1 t + a_1{}^2(vt, y) \cos[2\omega_1 t + 2\phi_1(vt, y)] \\ + 2a_0 a_1(vt, y) \cos[\phi_1(vt, y)] + 2a_0 a_1(vt, y) \cos[2\omega_1 t + \phi(vt, y)]\}$$

$$(4.14)$$

After low-pass filtering, the signal is

$$z(t) = \tfrac{1}{2}\{a_0{}^2 + a_1{}^2(vt, y) + 2a_0 a_1(vt, y) \cos[\phi_1(vt, y)]\} \qquad (4.15)$$

In this case, we again have a disturbing term.

Since many readers with an electronic background doubtless know of other methods for achieving the same result, we will not belabor this discussion.

4.2. SCANNED SOURCE

An interesting variation of holography appeared shortly after receiver scanning was developed. The suggestion was made that the role of receiver and source might be interchanged [1,2]. Inasmuch as source–receiver reciprocity in antenna theory exists, this idea is really not too surprising and we

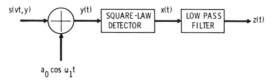

Fig. 4.4. The method of phase detection portrayed here is similar to the optical case and results in a hologram with a space-varying interference term.

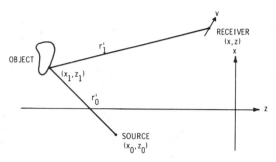

Fig. 4.5. Geometry for hologram recording with a
scanning receiver.

proceed to explore it with the aid of Fig. 4.5. We recall that the field distribution in the scanning plane is

$$s(x) = K \int \frac{O(x_1, z_1)}{r_0' r_1'(x)} \exp\{-ik_1[r_0' + r_1'(x)]\} \, dx_1 \, dz_1 \qquad (4.16)$$

When the receiver is scanned we obtain the time varying signal

$$s(t) = K \int \frac{O(x_1, z_1)}{r_0' r_1'(t)} \exp\{-ik_1[r_0' + r_1'(t)]\} \, dx_1 \, dz_1 \qquad (4.17)$$

Suppose now that we scan the source and leave the receiver fixed as shown
in Fig. 4.6. Then r_1' becomes the constant r_1'', and r_0' becomes the function
of time $r_0''(t)$ with the result that

$$s''(t) = K \int \frac{O(x_1, z_1)}{r_0''(t) r_1''} \exp\{-ik_1[r_0''(t) + r_1'']\} \, dx_1 \, dz_1 \qquad (4.18)$$

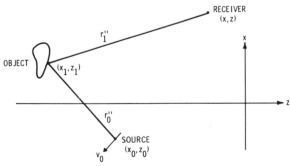

Fig. 4.6. Geometry for hologram recording with a scanning
source.

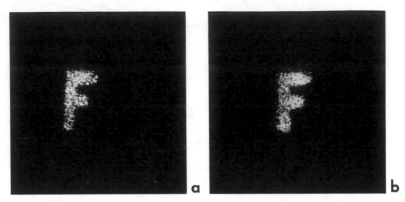

Fig. 4.7. These photographs illustrate images obtained from holograms made by (a) receiver scan and (b) source scan. In each case the geometry was the same but with source and receiver exchanged.

In order for expressions (4.17) and (4.18) to be equal, we have to satisfy the simultaneous equations

$$r_0' + r_1'(t) = r_0''(t) + r_1''$$
$$r_0'r_1'(t) = r_0''(t)r_1'' \tag{4.19}$$

Solving for $r_0''(t)$ and r_1'', we obtain

$$r_0''(t) = r_1'(t)$$
$$r_1'' - r_0' \tag{4.20}$$

In other words, the source and receiver are interchanged. The resulting holograms and images will be exactly equivalent as illustrated in Fig. 4.7.

4.3. SCANNED OBJECT

Having freed ourselves from the notion that only the receiver may be in motion, we ask whether it is possible for both the source and receiver to be stationary while the object is scanned? The answer is "yes," with some reservations. If the object is scanned as shown in Fig. 4.8, the signal from the receiver is

$$s'''(t) = K \int \frac{O(x_1, z_1)}{r_0'''(t)r_1'''(t)} \exp\{-ik_1[r_0'''(t) + r_1'''(t)]\} \, dx_1 \, dz_1 \tag{4.21}$$

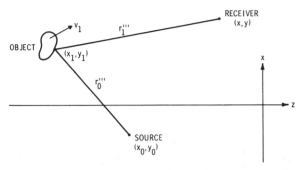

Fig. 4.8. Geometry for hologram recording with a scanning object.

The simultaneous equations to be solved in order to make $s(t) = s'''(t)$ are

$$r_0' + r_1'(t) = r_0'''(t) + r_1'''(t)$$
$$r_0'r_1'(t) = r_0'''(t)r_1'''(t)$$

(4.22)

The solutions are $r_1'''(t) = r_0'$, $r_0'''(t) = r_1'(t)$, or $r_1'''(t) = r_1'(t)$, $r_0'''(t) = r_0'$.
To fulfill the first solution set, we require that the object motion be an arc
about the receiver at a distance r_0'. This fulfills the condition $r_1'''(t) = r_0'$
and fixes the function $r_0'''(t)$, which in turn fixes the receiver scan motion
as shown in Fig. 4.9a. The equivalent object scan motion is shown in
Fig. 4.9b.

The other solution is attained when the object is scanned in an arc
about the source as shown in Fig. 4.9c. Note that the velocity of object
scan is different than that of receiver scan due to the different radius.

In all of these cases the hologram is scanned over a spherical surface
and to reconstruct properly the hologram would have to be written on a
similar surface. Normally a planar surface is desired. From Fig. 4.9 it is
obvious that this can be achieved by placing the center of the scanning arc

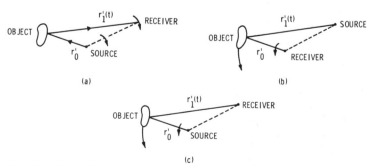

Fig. 4.9. These three scanning arrangements yield equivalent holograms.

at infinity. Then all arcs become planes and a planar hologram results. Figure 4.10 shows actual results equivalent to the three arrangements in Fig. 4.9, with the centers of scanning removed to infinity.

All of these geometries result in holograms exactly alike. In the next section we generalize to simultaneous source and receiver motion and derive

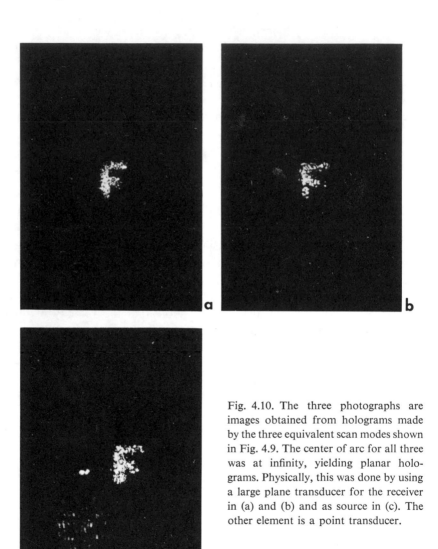

Fig. 4.10. The three photographs are images obtained from holograms made by the three equivalent scan modes shown in Fig. 4.9. The center of arc for all three was at infinity, yielding planar holograms. Physically, this was done by using a large plane transducer for the receiver in (a) and (b) and as source in (c). The other element is a point transducer.

the resulting images. The images are different from those obtained by receiver scanning but are nevertheless useful; in fact, they are superior for some applications.

4.4. SIMULTANEOUS SCANNING

Knowing that it is not necessary for certain parts of the system to remain stationary, we can now turn to the more general analysis for simultaneous source and receiver scanning [3]. Again, a precedent has been set in the field of radar theory [4]. In the Section 4.3 we came close to this concept by advocating object scanning. Since motion is relative, we could have kept the object stationary and scanned the source and receiver together.

The work in the previous section was concerned with making holograms identical with those obtained by receiver scanning. In this section we do not require identity; we take the approach of recording first and then asking questions about the image.

In this analysis we allow source and receiver to scan simultaneously. We can consider two types of scanning; one in which single units physically trace a path through space, or one in which arrays of sources and receivers are switched in sequence. It may be that the scanning is so slow, or the dwell time at each array element is so long, that the transit time of the radiation can be neglected. This is essentially what we have assumed up to this point. We will now, however, consider the case where transit time cannot be neglected.

We discovered in Chapter 2 that it was the phase of the radiation recorded in the hologram that determined the image position. Consequently, we consider only phase in this analysis. We will utilize the notation shown in Fig. 4.11 for recording and Fig. 4.12 for reconstruction of the hologram. The phase is recorded in the form of $\cos \phi(x)$ which can be decomposed into $\exp[i\phi(x)] + \exp[-i\phi(x)]$. Hence, we must consider $\pm \phi(x)$, which gives rise to the phenomenon of twin images.

The phase recorded at an arbitrary point as shown in Fig. 4.11 is

$$\phi_1 = -k_1(r_0' + r_1' - r_2') \tag{4.23}$$

This expression, however, assumes that the travel time of the radiation along the path $r_0' + r_1'$ is zero. In reality, the signal recorded at time t, was emitted by the source at time $t - \Delta t$, where Δt is the transit time $(r_0' + r_1')/c_1$ and c_1 is the velocity of propagation [5]. If, in the meantime, the source has moved, we must consider r_0' in Eq. (4.23) to be that distance

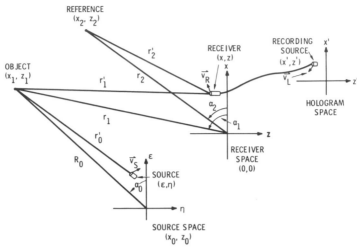

Fig. 4.11. Geometry and notation used for the analysis of simultaneous scanning of source and receiver.

at $t - \Delta t$. Let us then rewrite Eq. (4.23) to make this more apparent. Therefore,

$$\phi_1|_t = -k_1(r_0'|_{t-\Delta t} + r_1'|_t - r_2'|_t) \tag{4.24}$$

where $|_t$ represents a measurement at time t. We can approximate $r_0'|_{t-\Delta t}$ by

$$r_0'|_{t-\Delta t} = r_0'|_t - \Delta r_0' \tag{4.25}$$

where

$$\Delta r_0' = \frac{\partial r_0'}{\partial t} \Delta t$$

and

$$\Delta t = (r_0'|_t + r_1'|_t)/c_1$$

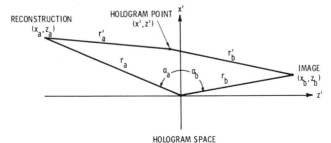

Fig. 4.12. Geometry and notation for image formation from a hologram.

This assumption is valid only if $\Delta r_0' \ll r_0' + r_1'$. Then the expression for the phase becomes

$$\phi_1 = -k_1(r_0' + r_1' - r_2') + \frac{k_1}{c_1} \frac{\partial r_0'}{\partial t}(r_0' + r_1') \qquad (4.26)$$

which is the same as Eq. (4.23) with the addition of a correction term. If we are physically scanning the source, $\partial r_0'/\partial t$ represents the component of velocity along the direction r_0'. If we are switching elements of an array, it represents the change in distance, r_0', in the switching interval, which can again be interpreted as a velocity component.

We expand the distance terms in a binomial series about the respective origins as follows and ignore the third and higher terms.

$$r_0' = R_0 + \frac{f_1(\varepsilon, \mu, \eta)}{2R_0} - \frac{[f_1(\varepsilon, \mu, \eta)]^2}{8R_0^3} + \cdots$$

$$r_1' = r_1 + \frac{f_2(x, y, z)}{2r_1} - \frac{[f_2(x, y, z)]^2}{8r_1^3} + \cdots \qquad (4.27)$$

$$r_2' = r_2 + \frac{f_3(x, y, z)}{2r_2} - \frac{[f_3(x, y, z)]^2}{8r_1^3} + \cdots$$

where

$$f_1(\varepsilon, \mu, \eta) = \varepsilon(t)[\varepsilon(t) - 2R_0 \cos \alpha_0] + \mu(t)[\mu(t) - 2R_0 \cos \beta_0]$$
$$+ \eta(t)[\eta(t) - 2R_0 \cos \gamma_0] \qquad (4.28)$$

$$f_2(x, y, z) = x(t)[x(t) - 2r_1 \cos \alpha_1] + y(t)[y(t) - 2r_1 \cos \beta_1]$$
$$+ z(t)[z(t) - 2r_1 \cos \gamma_1] \qquad (4.29)$$

$$f_3(x, y, z) = x(t)[x(t) - 2r_2 \cos \alpha_2] + y(t)[y(t) - 2r_2 \cos \beta_2]$$
$$+ z(t)[z(t) - 2r_2 \cos \gamma_2] \qquad (4.30)$$

The angles α, β, γ are measured from the positive x, y, z axes to the vector joining the origin to the point in question on the planes defined by the vector and the x, y, z axes, respectively. We see from Eq. (4.27) that we have included the time variable to account for the scanning action. From this point on we concern ourselves with two dimensions only, keeping in mind that all three are present and interact in a complex manner, especially when the third term of Eq. (4.27) is retained for aberration analysis. We will not retain it for the present. The final answers, however, will be given in three dimensions.

The first task we must perform is to relate the time varying values to velocities. As an example, consider $\varepsilon(t)$. This can be written as

$$\varepsilon(t) = \int_0^t v_\varepsilon(t) \, dt \tag{4.31}$$

where $v_\varepsilon(t)$ is the component of velocity of the source in the ε direction. We can obtain similar expressions for $\eta(t)$, $x(t)$, $z(t)$. Therefore, we find that the measured phase is a function of time. This electronic phase signal is used to modulate a light source which then scans a surface in the x', y', z' space (hologram space) at velocity $v_{x'}(t)$, $v_{y'}(t)$, $v_{z'}(t)$. Since $v_{x'}(t) = dx'(t)/dt$, we find

$$dt = \frac{dx'}{v_{x'}'} \tag{4.32}$$

Equation (4.31) can thus be converted from a function of time to a function of hologram space x' by substituting this value for dt. That is

$$\varepsilon(x') = \int_{\tilde{x}'}^{x'} \frac{v_\varepsilon(x')}{v_{x'}'(x')} \, dx' \tag{4.33}$$

where $\tilde{x}' = x'$ at $t = 0$. We expanded the distance r_0' in a binomial series and neglected all those terms of order higher than $f(\varepsilon, \mu, \eta)$ because the coefficients of ε and ε^2 enable us to find image location. The coefficients of $f^2(\varepsilon, \mu, \eta)$ are used to calculate aberrations. If we look at Eq. (4.33) we find that unless the integrand is independent of x', we will obtain terms of higher orders, which would then have to be included in the aberration terms. One can see that the problem could very quickly get out of hand. We require, therefore, that the velocity ratio $v_\varepsilon(t)/v_{x'}(t) = 1/m_1$, a constant. Therefore, we have

$$\varepsilon(x') = \frac{(x' - \tilde{x}')}{m_1} \tag{4.34}$$

The restriction placed on the velocity ratio is not too severe since it allows arbitrary velocity changes to occur as long as they occur proportionately in both.

Expressions (4.28) to (4.30) can then be written

$$f_1(x', y', z') = (x' - \tilde{x}')/m_1[(x' - \tilde{x}')/m_1 - 2R_0 \cos \alpha_0]$$
$$+ (y' - \tilde{y}')/m_2[(y' - \tilde{y}')/m_2 - 2R_0 \cos \beta_0]$$
$$+ (z' - \tilde{z}')/m_3[(z' - \tilde{z}')/m_3 - 2R_0 \cos \gamma_0] \tag{4.35}$$

$$f_2(x', y', z') = (x' - \tilde{x}')/m_4[(x' - \tilde{x}')/m_4 - 2r_1 \cos \alpha_1]$$
$$+ (y' - \tilde{y}')/m_5[(y' - \tilde{y}')/m_5 - 2r_1 \cos \beta_1]$$
$$+ (z' - \tilde{z}')/m_6[(z' - \tilde{z}')/m_6 - 2r_1 \cos \gamma_1] \qquad (4.36)$$

$$f_3(x', y', z') = (x' - \tilde{x}')/m_4[(x' - \tilde{x}')/m_4 - 2r_2 \cos \alpha_2]$$
$$+ (y' - \tilde{y}')/m_5(y' - \tilde{y}')/m_5 - 2r_2 \cos \beta_2]$$
$$+ (z' - \tilde{z}')/m_6[(z' - \tilde{z}')/m_6 - 2r_2 \cos \gamma_2] \qquad (4.37)$$

where

$$\frac{v_\mu}{v_y'} = \frac{1}{m_2}, \quad \frac{v_\eta}{v_z'} = \frac{1}{m_3}, \quad \frac{v_x}{v_x'} = \frac{1}{m_4}, \quad \frac{v_y}{v_y'} = \frac{1}{m_5}, \quad \frac{v_z}{v_z'} = \frac{1}{m_6}$$

4.4.1. Image Location

We now have the phase, as detected at the scanning mechanism, written on a diffracting medium in x', y', z' space. We illuminate this diffracting medium with a spherical light beam as shown in Fig. 4.12 with the result that the phase of the light as it leaves the medium is

$$\phi_2(x', y', z') = -k_2\left\{r_a + \frac{f_4(x', y', z')}{2r_a} - \frac{[f_4(x', y', z')]^2}{8r_a{}^3} + \cdots\right\}$$
$$\pm \phi_1(x', y', z') \qquad (4.38)$$

where

$$f_4(x', y', z') = x'(x' - 2r_a \cos \alpha_a) + y'(y' - 2r_a \cos \beta_a) + z'(z' - 2r_a \cos \gamma_a)$$

If ϕ_2 is to represent the phase due to another spherical wave focusing at a point x_b, y_b, z_b, it should equal

$$\phi_3(x', y', z') = -k_2\left\{r_b + \frac{f_5(x', y', z')}{2r_b} - \frac{[f_5(x', y', z')]^2}{8r_b{}^3} + \cdots\right\} \qquad (4.39)$$

where

$$f_5(x', y', z') = x'(x' - 2r_b \cos \alpha_b) + y'(y' - 2r_b \cos \beta_b) + z'(z' - 2r_b \cos \gamma_b)$$

The location of the Gaussian image point is obtained by equating the coefficients of x', y', z', $(x')^2$, $(y')^2$, $(z')^2$ of Eq. (4.39) to those of Eq. (4.38). This is relatively straightforward if transit time is neglected. If it is not, we have an error term as shown in Eq. (4.26). We will now proceed to evaluate it.

Let

$$\varepsilon_1 = \frac{k_1}{c_1} \frac{\partial r_0'}{\partial t} (r_0' + r_1') \tag{4.40}$$

Using the expression for r_0' as written in Eq. (4.27), we have

$$\frac{\partial r_0'}{\partial t} = \frac{1}{2R_0} \frac{\partial f_1}{\partial t} - \frac{f_1}{4R_0^3} \frac{\partial f_1}{\partial t} + \cdots$$

$$= \frac{1}{2R_0} \frac{\partial f_1}{\partial t} \left\{ 1 - \frac{f_1}{2R_0^2} + \cdots \right\}$$

$$= \frac{1}{2r_0'} \cdot \frac{\partial f_1}{\partial t} \tag{4.41}$$

Evaluating $\partial f_1 / \partial t$ gives

$$\frac{\partial f_1}{\partial t} = 2v_\varepsilon(t)[\varepsilon(t) - R_0 \cos \alpha_0] + 2v_\mu(t)[\mu(t) - R_0 \cos \beta_0]$$
$$+ 2v_\eta(t)[\eta(t) - R_0 \cos \gamma_0] \tag{4.42}$$

Converting this expression from time dependence to space dependence (x', y', z') yields

$$\frac{\partial f_1}{\partial t} = 2\{v_\varepsilon(x')[(x'-\tilde{x}')/m_1 - R_0 \cos \alpha_0] + v_\mu(y')[(y-\tilde{y}')/m_2 - R_0 \cos \beta_0]$$
$$+ v_\eta(z')[(z - \tilde{z}')/m_3 - R_0 \cos \gamma_0]\} \tag{4.43}$$

Thus, the error term becomes

$$\varepsilon_1 = \frac{1}{2} \frac{k_1}{c_1} \frac{\partial f_1}{\partial t} \left(\frac{r_0' + r_1'}{r_0'} \right) \tag{4.44}$$

We will now write down the complete expression for $\phi_2(x')$ using only one dimension so that we get an idea of the effect of transit time.

$$\phi_2(x') \cong -k_2 \left\{ r_a + \frac{x'(x' - 2r_a \cos \alpha_a)}{2r_a} \right\} \pm \phi_1(x') \tag{4.45}$$

where

$$\phi_1(x') = -k_1 \left\{ R_0 + \frac{(x' - \tilde{x}')/m_1[(x' - \tilde{x}')/m_1 - 2R_0 \cos \alpha_0]}{2R_0} \right.$$

$$+ r_1 + \frac{(x' - \tilde{x}')/m_4[(x' - \tilde{x}')/m_4 - 2r_1 \cos \alpha_1]}{2r_1}$$

$$\left. - r_2 - \frac{(x' - \tilde{x}')/m_4[(x' - \tilde{x}')/m_4 - 2r_2 \cos \alpha_2]}{2r_2} \right\}$$

$$+ k_1 \frac{v_\varepsilon(x')}{c_1} \left\{ \frac{(x' - \tilde{x}')/m_1 - R_0 \cos \alpha_0}{R_0} \right\}$$

$$\times \left\{ R_0 + \frac{(x' - \tilde{x}')/m_1[(x' - \tilde{x}')/m_1 - 2R_0 \cos \alpha_0]}{2R_0} \right.$$

$$\left. + r_1 + \frac{(x' - \tilde{x}')/m_4[(x' - \tilde{x}')/m_4 - 2r_1 \cos \alpha_1]}{2r_1} \right\}$$

This equation shows that the transit-time error has introduced a third order term in x', even though only the first-order term in f was retained in the binomial expansion. In addition, the factor $v_\varepsilon(x')/c_1$ is space variant unless $v_\varepsilon(x')$ is constant, which is not generally the case. Since the general function $v_\varepsilon(x')$ can be expanded in the power series

$$v_\varepsilon(x') = a_0 + a_1 x' + a_2 (x')^2 + \ldots$$

we see that $v_\varepsilon(x')$ introduces even more aberration terms. If, however, we assume slow velocity changes, $a_1, a_2, \ldots$ tend toward zero since they are derivatives of $v_\varepsilon(x')$. Then we can replace $v_\varepsilon(x')$ by its average, $\tilde{v}_\varepsilon$. If the source velocity is an appreciable fraction of the propagation velocity of the radiation used to make the hologram, we induce added aberration terms. The exact effect of this will be made clearer when image parameters are obtained.

If we also write down $\phi_3(x')$ in one dimension only, we can proceed to find the image location:

$$\phi_3(x') \cong -k_2 \left\{ r_b + \frac{x'(x' - 2r_b \cos \alpha_b)}{2r_b} \right\} \tag{4.46}$$

Equating coefficients of $(x')^2$ and x' of Eq. (4.46) and (4.45), we obtain

$$\frac{1}{r_b} = \pm \frac{k_1}{k_2} \left(\frac{1}{m_1^2 R_0} + \frac{1}{m_4^2 r_1} - \frac{1}{m_4^2 r_2} \right) - \frac{1}{r_a} \pm \left(\frac{v_\varepsilon}{c_1} \right) \varepsilon_2 \tag{4.47}$$

where

$$\varepsilon_2 = \frac{k_1}{k_2} \left\{ \frac{3 \cos \alpha_0}{m_1^2 R_0} + \left(\frac{2 \cos \alpha_1}{m_1 m_4} + \frac{\cos \alpha_0}{m_4^2} \right) \frac{1}{r_1} \right\}$$

and

$$\cos \alpha_b = \pm \frac{k_1}{k_2} \left\{ \left(\frac{\cos \alpha_0}{m_1} + \frac{\cos \alpha_1}{m_4} - \frac{\cos \alpha_2}{m_4} \right) \right.$$

$$\left. + \tilde{x}' \left(\frac{1}{m_1^2 R_0} + \frac{1}{m_4^2 r_1} - \frac{1}{m_4^2 r_2} \right) \right\} - \cos \alpha_a \pm \frac{\tilde{v}_\varepsilon}{c_1} \varepsilon_3 \tag{4.48}$$

where

$$\varepsilon_3 = \frac{k_1}{k_2} \left\{ \frac{1}{m_1} \left(1 + \frac{r_1}{R_0} + \cos^2 \alpha_0 \right) + \frac{\cos \alpha_1 \cos \alpha_0}{m_4} \right\}$$

If we include all three dimensions we obtain two more sets of equations; namely,

$$\frac{1}{r_b} = \pm \frac{k_1}{k_2} \left(\frac{1}{m_2^2 R_0} + \frac{1}{m_5^2 r_1} - \frac{1}{m_5^2 r_2} \right) - \frac{1}{r_a} \pm \frac{\tilde{v}_\mu}{c_1} \varepsilon_4 \quad (4.49)$$

where

$$\varepsilon_4 = \frac{k_1}{k_2} \left\{ \frac{3 \cos \beta_0}{m_2^2 R_0} + \left(\frac{2 \cos \beta_1}{m_2 m_5} + \frac{\cos \beta_0}{m_5^2} \right) \frac{1}{r_1} \right\}$$

$$\cos \beta_b = \pm \frac{k_1}{k_2} \left\{ \left(\frac{\cos \beta_0}{m_2} + \frac{\cos \beta_1}{m_5} - \frac{\cos \beta_2}{m_5} \right) \right.$$

$$\left. + \tilde{y}' \left(\frac{1}{m_2^2 R_0} + \frac{1}{m_5^2 r_1} - \frac{1}{m_5^2 r_2} \right) \right\} - \cos \beta_a \pm \frac{\tilde{v}_\mu}{c_1} \varepsilon_5 \quad (4.50)$$

where

$$\varepsilon_5 = \frac{k_1}{k_2} \left\{ \frac{1}{m_2} \left(1 + \frac{r_1}{R_0} + \cos^2 \beta_0 \right) + \frac{\cos \beta_1 \cos \beta_0}{m_5} \right\}$$

and

$$\frac{1}{r_b} = \pm \frac{k_1}{k_2} \left(\frac{1}{m_3^2 R_0} + \frac{1}{m_6^2 r_1} - \frac{1}{m_6^2 r_2} \right) - \frac{1}{r_a} \pm \frac{\tilde{v}_\eta}{c_1} \varepsilon_6 \quad (4.51)$$

where

$$\varepsilon_6 = \frac{k_1}{k_2} \left\{ \frac{3 \cos \gamma_0}{m_3^2 R_0} + \left(\frac{2 \cos \gamma_1}{m_3 m_6} + \frac{\cos \gamma_0}{m_6^2} \right) \frac{1}{r_1} \right\}$$

$$\cos \gamma_b = \pm \frac{k_1}{k} \left\{ \left(\frac{\cos \gamma_0}{m_3} + \frac{\cos \gamma_1}{m_6} - \frac{\cos \gamma_2}{m_6} \right) \right.$$

$$\left. + \tilde{z}' \left(\frac{1}{m_3^2 R_0} + \frac{1}{m_6^2 r_1} - \frac{1}{m_6^2 r_2} \right) \right\} - \cos \gamma_a \pm \frac{v_\eta}{c_1} \varepsilon_7 \quad (4.52)$$

where

$$\varepsilon_7 = \frac{k_1}{k_2} \left\{ \frac{1}{m_3} \left(1 + \frac{r_1}{R_0} + \cos^2 \gamma_0 \right) + \frac{\cos \gamma_1 \cos \gamma_0}{m_6} \right\}$$

Note that we now have three different equations for the radial distance to the image point. This is known as astigmatism and is something that can be corrected by astigmatic optics; however, in general, we wish to provide a stigmatic image. This requires that all three equations for r_b be equal. If transit time can be neglected, all that is required is that $m_1 = m_2 = m_3 = m_S$ and $m_4 = m_5 = m_6 = m_R$. This means that the vector velocities are all in the same direction with magnitudes related by $|\mathbf{v}_L|/|\mathbf{v}_S| = m_S$, $|\mathbf{v}_L|/|\mathbf{v}_R|$

$| = m_R$, where $|\mathbf{v}_L|, |\mathbf{v}_S|, |\mathbf{v}_R|$ are the magnitudes of the velocities of the recording light source, the radiation source, and the receiver, respectively.

If transit time must be considered, a comparison of the error terms of the three equations for image distance reveals that the errors are not equal even when the velocity ratios are corrected, except for $\alpha_0 = \beta_0 = \gamma_0$ and $\alpha_1 = \beta_1 = \gamma_1$. Thus, we might say that we have some astigmatism introduced when fast source scanning is used.

We now write down the image equations with the following conditions: (1) $|\tilde{\mathbf{v}}_\varepsilon| \ll c_1$ and (2) $\tilde{x}' = \tilde{y}' = \tilde{z}' = 0$. The last condition is merely a shifting of the hologram coordinate system. Then

$$\frac{1}{r_b} = \pm \frac{k_1}{k_2} \left(\frac{1}{m_S^2 R_0} + \frac{1}{m_R^2 r_1} - \frac{1}{m_R^2 r_2} \right) - \frac{1}{r_a} \qquad (4.53)$$

$$\cos \alpha_b = \pm \frac{k_1}{k_2} \left(\frac{\cos \alpha_0}{m_S} + \frac{\cos \alpha_1}{m_R} - \frac{\cos \alpha_2}{m_R} \right) - \cos \alpha_a \qquad (4.54)$$

$$\cos \beta_b = \pm \frac{k_1}{k_2} \left(\frac{\cos \beta_0}{m_S} + \frac{\cos \beta_1}{m_R} - \frac{\cos \beta_2}{m_R} \right) - \cos \beta_a \qquad (4.55)$$

The equation for $\cos \gamma_b$ is superfluous and will not be used.

We will now check these equations against those we derived in a less general manner. The first case is that of a stationary source ($\mathbf{v}_S = 0$). Then we have

$$\frac{1}{r_b} = \pm \frac{k_1}{m_R^2 k_2} \left(\frac{1}{r_1} - \frac{1}{r_2} \right) - \frac{1}{r_a}$$

$$\cos \alpha_b = \pm \frac{k_1}{m_R k_2} (\cos \alpha_1 - \cos \alpha_2) - \cos \alpha_a \qquad (4.56)$$

$$\cos \beta_b = \pm \frac{k_1}{m_R k_2} (\cos \beta_1 - \cos \beta_2) - \cos \beta_a$$

These equations are identical to those derived in Chapter 2. The ratio, $m_R = |\mathbf{v}_L| / |\mathbf{v}_R|$ plays the role of the photographic enlargement factor in optical holography.

The second case we wish to test is that of stationary receiver ($\mathbf{v}_R = 0$). Then we have

$$\frac{1}{r_b} = \pm \frac{k_1}{m_S^2 k_2} \frac{1}{R_0} - \frac{1}{r_a}$$

$$\cos \alpha_b = \pm \frac{k_1}{m_S k_2} \cos \alpha_0 - \cos \alpha_a \qquad (4.57)$$

$$\cos \beta_b = \pm \frac{k_1}{m_S k_2} \cos \beta_0 - \cos \beta_a$$

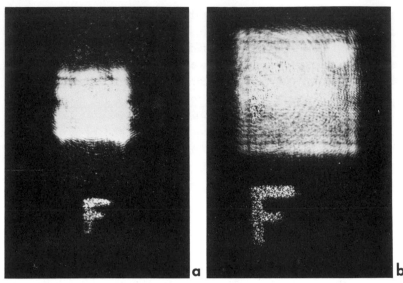

Fig. 4.13. These photographs show (a) the image from a receiver scan and (b) a source–receiver scan hologram. Note that the latter is twice as large as the former due to the fact that it appears twice as close to the hologram. Since the photographs were taken the same distance from the hologram, the latter image is enlarged.

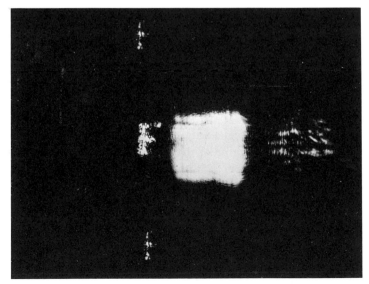

Fig. 4.14. This image was taken from a Gabor hologram made with a plane source and point receiver scanned together. The source was tilted at an angle with respect to the object. Note that image offset has been achieved.

This set of equations shows that the image appears as seen through the source scan aperture, which is intuitively satisfying.

A special case of interest is one in which a single transducer acts as both source and receiver in the manner of a radar or sonar system. Then $m_S = m_R = m$, $R_0 = r_1$, $\alpha_0 = \alpha_1$, $\beta_0 = \beta_1$ and

$$\frac{1}{r_b} = \pm \frac{k_1}{m^2 k_2} \left(\frac{2}{r_1} - \frac{1}{r_2} \right) - \frac{1}{r_a}$$

$$\cos \alpha_b = \pm \frac{k_1}{mk_2} (2 \cos \alpha_1 - \cos \alpha_2) - \cos \alpha_a \qquad (4.58)$$

$$\cos \beta_b = \pm \frac{k_1}{mk_2} (2 \cos \beta_1 - \cos \beta_2) - \cos \beta_a$$

The most obvious effect of this particular scheme is to bring the image closer, as evidenced by the terms $2/r_1$ and $2 \cos \alpha_1$. More will be said about this particular arrangement later.

A second special case is one in which both α_1 and α_2 are $90°$. The resulting image position is

$$\frac{1}{r_b} = \pm \frac{k_1}{k_2} \left(\frac{1}{m_S^2 R_0} + \frac{1}{m_R^2 r_1} - \frac{1}{m_R^2 r_2} \right) - \frac{1}{r_a}$$

$$\cos \alpha_b = \pm \frac{k_1}{m_S k_2} \cos \alpha_0 - \cos \alpha_a \qquad (4.59)$$

$$\cos \beta_b = \pm \frac{k_1}{m_S k_2} \cos \beta_0 - \cos \beta_a$$

The point of interest here is that, although we have made a Gabor hologram (object and reference on-axis), the image is still off-axis. The image distance, however, remains unchanged.

Figures 4.13 and 4.14 illustrate the results of the simultaneous scan system and the offset-Gabor hologram.

4.4.2. Resolution

An important consideration in any imaging system is resolution. So far we have found that it is always dependent upon the distance from the aperture to the object point and inversely upon the aperture size. We defined resolution on the basis of the number of fringes in the Fresnel zone pattern within the aperture. It is difficult to define "aperture" when the hologram surface is other than a plane. Consequently, we restrict ourselves to a plane hologram, which means that the receiver and source scanning surfaces are also planar. If we restrict the hologram aperture to be A, centered at $x' = 0$, we find from Eq. (4.45) that the phase difference between the aperture extremes is

$$\phi = \phi_1(A/2) - \phi_1(-A/2) = k_1 A \left\{ \frac{\cos \alpha_0}{m_S} + \frac{\cos \alpha_1}{m_R} - \frac{\cos \alpha_2}{m_R} \right\} - \frac{\tilde{v}_\varepsilon}{c} \varepsilon_8$$

(4.60)

where

$$\varepsilon_8 = k_1 A \left\{ \frac{1}{m_S} \left(1 + \frac{r_1}{R_0} + \cos^2 \alpha_0 \right) + \frac{\cos \alpha_1 \cos \alpha_0}{m_R} \right\}$$

The incremental change in phase for an incremental change in object angular position is

$$\Delta\phi = \frac{\partial \phi}{\partial \alpha_1} \Delta\alpha_1 = k_1 A \, \Delta\alpha_1 \left\{ \left[\frac{\sin \alpha_0}{m_S} \frac{\alpha_0}{\alpha_1} + \frac{\sin \alpha_1}{m_R} \right] - \frac{v_\varepsilon}{c_1} \varepsilon_9 \right\}$$

(4.61)

where

$$\varepsilon_9 = \frac{1}{m_S} \left(\frac{r_1}{R_0^2} \frac{\partial R_0}{\partial \alpha_1} + 2\cos \alpha_0 \sin \alpha_1 \frac{\partial \alpha_0}{\partial \alpha_1} \right)$$

$$+ \frac{1}{m_R} \left(\sin \alpha_1 \cos \alpha_0 + \cos \alpha_1 \sin \alpha_0 \frac{\partial \alpha_0}{\partial \alpha_1} \right)$$

By simple geometrical arguments

$$\frac{\partial \alpha_0}{\partial \alpha_1} = \frac{r_1}{R_0} \cos(\alpha_1 - \alpha_0), \quad \frac{\partial R_0}{\partial \alpha_1} = r_1 \sin(\alpha_1 - \alpha_0)$$

(4.62)

We substitute Eq. (4.62) into Eq. (4.61), set it equal to 2π and solve for $\Delta\alpha_1$ to obtain the angular resolution

$$\Delta\alpha_1 = \frac{\lambda_1}{A \left\{ \frac{r_1}{m_S R_0} \sin \alpha_0 \cos(\alpha_1 - \alpha_0) + \frac{1}{m_R} \sin \alpha_1 - \frac{\tilde{v}_\varepsilon}{c_1} \varepsilon_{10} \right\}}$$

(4.63)

where

$$\varepsilon_{10} = \frac{r_1}{m_S R_0} \left[\frac{r_1}{R_0} \sin(\alpha_1 - \alpha_0) + 2\cos \alpha_0 \sin \alpha_0 \cos(\alpha_1 - \alpha_0) \right]$$

$$+ \frac{1}{m_R} \left[\sin \alpha_1 \cos \alpha_0 + \frac{r_1}{R_0} \cos \alpha_1 \sin \alpha_0 \cos(\alpha_1 - \alpha_0) \right]$$

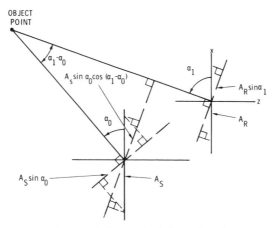

Fig. 4.15. Geometry for resolution calculations for the source–receiver scanned hologram.

This equation is easily interpreted with the help of Fig. 4.15. First we must recognize that $A/m_S = A|\mathbf{v}_S|/|\mathbf{v}_L| = A_S$, the source scan aperture and $A/m_R = A_R$, the receiver scan aperture. The term, $(A/m_S)\sin\alpha_0\cos(\alpha_1 - \alpha_0)$ $= A_S\sin\alpha_0\cos(\alpha_1 - \alpha_0)$, then represents the source scan aperture projected onto a plane perpendicular to the vector $\mathbf{r}_1$ and $(A/m_R)\sin\alpha_1 = A_R\sin\alpha_1$ is the projection of the receiver scan aperture on a plane perpendicular to the same vector. Rewriting Equation (4.63) we have

$$\Delta\alpha_1 = \frac{\lambda_1}{\left\{A\dfrac{r_1}{R_0}\sin\alpha_0\cos(\alpha_1 - \alpha_0) + A_R\sin\alpha_1\right\} - \dfrac{\tilde{v}_\varepsilon}{c_1}\varepsilon_{11}} \qquad (4.64)$$

where

$$\varepsilon_{11} = A_S\frac{r_1}{R_0}\left[\frac{r_1}{R_0}\sin(\alpha_1 - \alpha_0) + \sin 2\alpha_0\cos(\alpha_1 - \alpha_0)\right]$$

$$+ A_R\left[\sin\alpha_1\cos\alpha_0 + \frac{r_1}{R_0}\cos\alpha_1\sin\alpha_0\cos(\alpha_1 - \alpha_0)\right]$$

To attain maximum resolution we must maximize the denominator of Eq. (4.64) for all α_1. If we assume $\tilde{v}_\varepsilon \ll c_1$, this is accomplished by orienting the scanning planes so that $\alpha_0 = \alpha_1$. This can only be achieved by having the scanning planes coincident, with the result that

$$\Delta\alpha_1 = \frac{\lambda_1}{(A_S + A_R)\sin\alpha_1} \qquad (4.65)$$

The error term ε_{11} for this choice of geometry becomes

$$\varepsilon_{11} = (A_S + A_R) \sin 2\alpha_1 \qquad (4.66)$$

We see that the resolution is maximum and the error zero for $\alpha_1 = 90°$. Since an object always has a range of angles, there is always some reduction of resolution due to transit time effects.

We now wish to explore the resolution as a function of scanning velocities. We can write

$$A_S + A_R = A_S\left(1 + \frac{A_R}{A_S}\right)$$

or

$$A_S + A_R = A_R\left(1 + \frac{A_S}{A_R}\right)$$

If we are limited to a maximum aperture, $\tilde{A}$, for either the source or receiver scans, we have

$$A_S + A_R = \tilde{A}\left(1 + \frac{A_R}{\tilde{A}}\right) \qquad \text{for } A_R \leq A_S$$

or

$$A_S + A_R = \tilde{A}\left(1 + \frac{A_S}{\tilde{A}}\right) \qquad \text{for } A_R \geq A_S$$

Obviously the largest value of $A_S + A_R$ under this restriction occurs when $A_R = A_S = \tilde{A}$. This, in turn, means that $m_R = m_S = m$ and since $m_S/m_R = v_R/v_S = p = 1$ we have source and receiver scan velocities equal. The optimum resolution then, is

$$\varDelta\alpha_1 = \frac{\lambda_1}{2\tilde{A}\left(\sin \alpha_1 - \dfrac{\tilde{v}_\varepsilon}{c_1} \sin 2\alpha_1\right)} \qquad (4.67)$$

If $\tilde{v}_\varepsilon \ll c_1$, this expression shows that resolution for the simultaneous scan system is twice that for the single scan system. This fact has long been known to those people working with high resolution coherent radars [3].

4.4.3. Image Location Revisited

It is now appropriate to return to the image location equations with our knowledge of the optimum system with respect to resolution. We consider the coincident scanning case where all scanning surfaces are planes.

Then

$$\frac{1}{r_b} = \left\{\pm \frac{k_1}{m^2 k_2}\left(\frac{2}{r_1} - \frac{1}{r_2}\right) - \frac{1}{r_a}\right\} \pm \left\{6\frac{k_1}{m^2 k_2}\frac{\tilde{v}_\varepsilon}{c_1}\cos\alpha_1\right\} \quad (4.68)$$

$$\cos\alpha_b = \left\{\pm \frac{k_1}{mk_2}(2\cos\alpha_1 - \cos\alpha_2) - \cos\alpha_a\right\} \pm \left\{2\frac{k_1}{mk_2}\frac{\tilde{v}_\varepsilon}{c_1}(1 + \cos^2\alpha_1)\right\}$$

$$(4.69)$$

where

$$m = m_S = m_R$$

If we first assume $\tilde{v}_\varepsilon \ll c_1$ and $r_a = r_2 m^2 k_2/k_1$, $\cos\alpha_a = (k_1/mk_2)\cos\alpha_2$ and consider the true image, we obtain

$$r_b = -\frac{m^2 k_2}{k_1}\left(\frac{r_1}{2}\right) \quad (4.70)$$

$$\cos\alpha_b = -2\frac{k_1}{mk_2}\cos\alpha_1 \quad (4.71)$$

The first equation shows that the image distance has been reduced by two. This can be quite an advantage when $k_2/k_1 = \lambda_1/\lambda_2$ is as large as it usually is for acoustical holography. The second equation, however, sets an angular limitation on the placement of the object (α_1). Since $|\cos\alpha_b| \leq 1$, we have $|\cos\alpha_1| < mk_2/2k_1 = m\lambda_1/2\lambda_2$; whereas, for receiver scan only the limitation would be $|\cos\alpha_1| \leq m\lambda_1/\lambda_2$.

We demonstate the effect of simultaneous scan with and without transit time consideration by using a plane object. Images for $\tilde{v}_\varepsilon/c_1 = 0$, and 0.1 are plotted along with the image for receiver scan only. We use light only

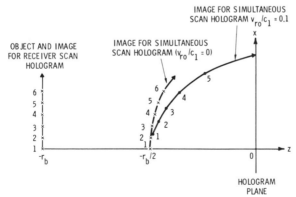

Fig. 4.16. Distortions introduced in a hologram due to a large source scan velocity.

(i.e., $k_1 = k_2$) and $m = 1$. The results are shown in Fig. 4.16. Note that when transit time is negligible, the image falls on a sphere of radius $r_1/2$. This is normal for any image not exactly reproducing the object (in this case, its distance from the hologram is different). We in our terminology, would consider the image aberration free. When transit time is taken into account the image is distorted. First, an angular shift has occurred and second, the image does not fall on a sphere.

We should point out that if we consider a λ_2/λ_1 ratio typical of acoustical holography, the distortion effect is so small as to be negligible.

4.4.4. Magnification

We defined the various magnifications to be:

Radial $\qquad M_{r_1} = \dfrac{\partial r_b}{\partial r_1} = \pm \dfrac{k_1}{k_2} \left(\dfrac{r_b}{r_1} \right)^2 \left\{ \dfrac{1}{m_R{}^2} + \left(\dfrac{r_1}{m_S R_0} \right)^2 \cos(\alpha_1 - \alpha_0) \right\}$

$$(4.72)$$

Angular $\qquad M_\alpha = \dfrac{\partial a_b}{\partial \alpha_1} = \pm \dfrac{k_1}{k_2} \left(\dfrac{\sin \alpha_1}{\sin \alpha_b} \right) \left\{ \dfrac{1}{m_R} + \left(\dfrac{r_1}{m_S R_0} \right) \dfrac{\sin \alpha_0}{\sin \alpha_1} \right\}$

$$(4.73)$$

Normal $\qquad M_n = \dfrac{r_b}{r_1} \dfrac{\partial \alpha_b}{\partial \alpha_1}$ $\hfill (4.74)$

where we have assumed $\tilde{v}_\varepsilon \ll c_1$. For the special case of coincident scanning planes ($\alpha_0 = \alpha_1$, $R_0 = r_1$, $m_S = m_R = m$) we have

$$M_R = \pm 2 \frac{k_1}{m^2 k_2} \left(\frac{r_b}{r_1} \right)^2 \qquad (4.75)$$

$$M_\alpha = \pm 2 \frac{k_1}{m k_2} \frac{\sin \alpha_1}{\sin \alpha_b} \qquad (4.76)$$

4.4.5. Aberrations

The third-order or Seidel aberrations are obtained by retaining the third term in the binomial series expansions of the distance terms, as shown in Eq. (4.27). Obviously, if this is done for all distance terms and transit time is considered, the expressions become very complicated. Hence, we confine ourselves to the special case for which transit time can be ignored [4]. For this case then, we have the aberrations:

Spherical
$$S = \pm \frac{k_1}{k_2} \left(\frac{1}{m_R{}^4 r_1{}^3} + \frac{1}{m_S{}^4 R_0{}^3} - \frac{1}{m_R{}^4 r_2{}^3} \right) + \frac{1}{r_a{}^3} + \frac{1}{r_b{}^3} \qquad (4.77)$$

Coma

$$C_x = \pm \frac{k_1}{k_2} \left(\frac{x_1}{m_R{}^3 r_1{}^3} + \frac{x_0}{m_S{}^3 R_0{}^3} - \frac{x_2}{m_R{}^3 r_2{}^3} \right) + \frac{x_a}{r_a{}^3} + \frac{x_b}{r_b{}^3} \qquad (4.78)$$

$$C_y = \pm \frac{k_1}{k_2} \left(\frac{y_1}{m_R{}^3 r_1{}^3} + \frac{y_0}{m_S{}^3 R_0{}^3} - \frac{y_2}{m_R{}^3 r_2{}^3} \right) + \frac{y_a}{r_a{}^3} + \frac{y_b}{r_b{}^3}$$

Astigmatism

$$A_x = \pm \frac{k_1}{k_2} \left(\frac{x_1{}^2}{m_R{}^2 r_1{}^3} + \frac{x_0{}^2}{m_S{}^2 R_0{}^3} - \frac{x_2{}^2}{m_R{}^2 r_2{}^3} \right) + \frac{x_a{}^2}{r_a{}^3} + \frac{x_b{}^2}{r_b{}^3} \qquad (4.79)$$

$$A_y = \pm \frac{k_1}{k_2} \left(\frac{y_1{}^2}{m_R{}^2 r_1{}^3} + \frac{y_0{}^2}{m_S{}^2 R_0{}^3} - \frac{y_2{}^2}{m_R{}^2 r_2{}^3} \right) + \frac{y_a{}^2}{r_a{}^3} + \frac{y_b{}^2}{r_b{}^3}$$

Field curvature

$$A_{xy} = \pm \frac{k_1}{k_2} \left(\frac{x_1 y_1}{m_R{}^2 r_1{}^3} + \frac{x_0 y_0}{m_S{}^2 R_0{}^3} - \frac{x_2 y_2}{m_R{}^2 r_2{}^3} \right) + \frac{x_a y_a}{r_a{}^3} + \frac{x_b y_b}{r_b{}^3} \qquad (4.80)$$

All aberrations disappear for only three cases, (1) stationary source; (2) stationary receiver; and (3) coincident source and receiver. The reason for moving the source space to coincide with the origin ($\alpha_0 = \alpha_1$, $R_0 = r_1$) in the stationary receiver system is that the magnification expressions are derived with respect to r_1, α_1. If the source aperture is not centered at the origin, we have a term of the type $\cos(\alpha_1 - \alpha_0)$ as a multiplier in the magnification expressions. In reality we want magnification with respect to R_0, which is equivalent to shifting the coordinate system to coincide with the center of the source aperture. Then $\cos(\alpha_1 - \alpha_0) = 1$ and our expressions are correct.

4.4.6. Distortion

As discussed in Section 2.9, distortion refers to the different radial and normal magnifications suffered when images are produced by any optical system. From the magnification Eqs. (4.72) to (4.74) we can relate the radial to normal magnification. We do this for the special case of coincident scanning planes (but not equal scan velocities) and $\tilde{v}_\varepsilon \ll c_1$. Then we obtain

$$M_r = \pm \left(\frac{k_2}{k_1} \right) \left(\frac{\sin \alpha_b}{\sin \alpha_1} \right)^2 \left[\frac{1 + p^2}{(1 + p)^2} \right] M_n{}^2 \qquad (4.81)$$

where

$$p = \frac{m_R}{m_S} = \frac{v_S}{v_R}$$

For reference we rewrite this relationship when only receiver scanning is used

$$M_r = \pm \frac{k_2}{k_1} \left(\frac{\sin \alpha_b}{\sin \alpha_1} \right)^2 M_n{}^2 \qquad (4.82)$$

Note that in the latter case there is no control on the distortion. In the former, however, we can control the ratio of source and receiver scan velocities to obtain undistorted magnification. We now proceed to discover the limitations to this method of defeating distortion. For simplicity we consider the paraxial approximation $(\alpha_b = \alpha_1 = 90°)$. Then setting $M_r = M_n = M$, we find that

$$\frac{1 + p^2}{(1 + p)^2} = \frac{1}{\mu M} = \frac{1}{K} \qquad (4.83)$$

where

$$\mu = k_2/k_1$$

and

$$K = \mu M$$

Solving for p, we have

$$p = \frac{-1}{K - 1} \pm \sqrt{\left(\frac{1}{K - 1} \right)^2 - 1} \qquad (4.84)$$

Since p must be real, we have the inequality $| K - 1 | \leq 1$ or

$$0 \leq M \leq 2 \frac{k_1}{k_2} = 2 \frac{\lambda_2}{\lambda_1} \qquad (4.85)$$

These extremes of magnification are obtained when $p = -1$ for $M = 0$ and $p = +1$ for $M = 2k_1/k_2$.

This discussion has shown that the promise of undistorted magnification indicated by Eq. (4.81) is somewhat illusory. It is useful only if k_1/k_2 is large, or if demagnification is desired. In acoustical holography k_1/k_2 is normally small.

4.5. RECORDING AND RECONSTRUCTION

As we mentioned earlier, the signal may be recorded on magnetic tape for storage, but eventually the information must be presented to the observer in the form of an image.

If the data were recorded as a Fourier transform hologram, then the image could be retrieved as a computer print-out, simply by having the computer perform an inverse Fourier transform on the recorded data [6]. Naturally, this can only work for two-dimensional objects. The reason for this is that any of the methods for taking the Fourier transform of the object field are two-dimensional in nature. In addition, any computer print-out has to be two-dimensional.

The only way to get the full amount of information from the data is to reconstruct the wave front. To do this we must record the data in the form of a hologram and then illuminate it with light. The basic method is to intensity-modulate a light source with the detected signal about some average intensity, move it in position and velocity in proportion to that of the detector, and record the light on a light-sensitive medium that is shaped proportionately to the surface over which the receiver was scanned.

In practice only planar surfaces have been used, but scan lines in the surfaces have been straight lines or circles. The most common light sources have been an incandescent bulb focused to a very small point, and the face of a cathode-ray tube with a moving intensity modulated spot. The first method requires a mechanical scanning system, most often the same one used for the source/receiver, and the second requires an electronic scanning system driven by electrical pickoffs on the source/receiver mechanism. Each method requires a camera to photograph the moving light source.

The cathode-ray method holds the potential for real-time (no photographic delay) image display if a special face is used. Such a system has been reported for optical filtering operations [7] and will be described in detail in Chapter 7.

We have been able, using a circular scanning system and electronic cathode-ray recording, to record an acoustical hologram in 5 sec. The photographic development time using Polaroid film was 10 sec for a total elapsed time of 15 sec.

4.6. TIME GATING

In some applications of acoustical holography, it is advantageous to use an auxiliary means to reject data from particular regions of object space. A graphic example of such a situation is the problem of finding flaws in a large casting. In such a case, the major part of the signal will come from reflection at the surface and only a small part from the internal flaw. The image from a hologram made with such a signal will generally not show the flaw at all.

This problem can be solved by using pulsed sound. An electronic gate, interposed between the receiver and phase detector, is enabled after a certain time delay. This has the effect of ignoring information coming from a region in object space

$$\Delta r = c_1 \Delta t$$

where Δt is the delay time. Thus, the front surface reflection can be eliminated from the hologram. Furthermore, by keeping the gate open a fixed time, T, it is possible to also ignore information from, for example, the back surface. That is, we can make a hologram of a thin slice of the object and the particular slice is chosen by sliding the gate pulse in time.

In practice, we use stationary source and receiver to echo-range the flaw, by watching the return signal on an oscilloscope while the gate is ranged through the object. When we see a large signal return that might be coming from a flaw, we make a hologram at this range. The image then enables us to see the exact size and shape of the flaw.

REFERENCES

1. A. F. Metherell and S. Spinak, Acoustical holography of nonexistent wavefronts detected at a single point in space, *Appl. Phys. Lett.* **13**:22 (1968).
2. V. I. Neeley, Source Scanning holography, *Phys. Lett.* **28A**(7):475–476 (1968).
3. B. P. Hildebrand and K. A. Haines, Holography by scanning, *J. Opt. Soc. Am.* **59**(1):1 (1969).
4. L. J. Cutrona, E. N. Leith, L. J. Porcello, and W. E. Vivian, On the application of coherent optical processing techniques to synthetic-aperture radar, *Proc. IEEE* **54**: 1026 (1966).
5. B. P. Hildebrand, The effect of high scanning velocities on the holographic image, *J. Opt. Soc. Am.* **60**(9):1166 (1970).
6. J. W. Goodman and R. W. Lawrence, Digital image formation from electronically detected holograms, *App. Phys. Lett.* **11**(3):77 (1967).
7. D. Casasent, An on-line optical data processing system, *Proceedings of the Electro-Optical Systems Design Conference*, New York City, Ed., K. A. Koptetzky, available from Industrial and Scientific Conference Management Inc. 222 West Adams St., Chicago, Ill., page 56 (1969).

Chapter 5

Sampled Holograms

In the final analysis, all recording devices are sampling devices, simply because they are imperfect. An acoustic receiver cannot be made infinitely small or be capable of following all rates of amplitude variation; film cannot be made with infinitely small grain size. Therefore, any detection and recording system can be considered to be a sampling system.

5.1. SAMPLING THEORY

Fortunately, there exists a theorem called the "sampling theorem," which tells us that not all is lost [1]. This theorem states that any wave front extending over any aperture, A, containing spatial variations limited to B lines/cm can be completely specified if we sample its amplitude at intervals of $1/2B$ cm [2]. It follows that for a square aperture the wave front is completely specified by $4B^2A^2$ samples. Thus, if our perfect point receiver is moved to $4B^2A^2$ positions or an array of $4B^2A^2$ receivers are used, sufficient information is gathered to completely reconstruct the wave front within the aperture. If this number of samples is not obtained, large errors in the reconstruction occur [3]. The way in which these errors occur and the resulting effect on the image will be discussed later.

In the following analysis we will use the properties of Dirac delta functions, Fourier series, and Fourier transforms in both one- and two-dimensions. We use the terminology of crystallography rather than electronics, since we are dealing with space and not with time.

5.1.1. One-Dimensional Sampling Theory

The basic tool in the analysis of the sampling process is the sampling function. This function is the uniform periodic sequence of delta functions in the direct space t, written as

$$s(t) = \sum_{m=-\infty}^{\infty} \delta(t - mT_0) \qquad (5.1)$$

where T_0 is the period.

The frequency spectrum in reciprocal space ω, of this sampling function, found by taking the Fourier transform of Eq. (5.1) is

$$S(\omega) = \sum_{m=-\infty}^{\infty} \exp(-i\omega mT_0) \qquad (5.2)$$

$$= \omega_0 \sum_{m=-\infty}^{\infty} \delta(\omega - m\omega_0) \qquad (5.3)$$

where $\omega_0 = 2\pi/T_0$. The reader may check this last result by writing it as a Fourier series. Thus, a periodic array of delta functions in direct space is also an array of delta functions in reciprocal space, as shown in Fig. 5.1.

The usefulness of the sampling function arises because of a property of delta functions; namely, that multiplication of an arbitrary function $f(t)$

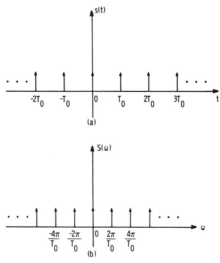

Fig. 5.1. (a) The periodic one-dimensional sampling function, $s(t)$ and (b) its Fourier transform, $S(\omega)$.

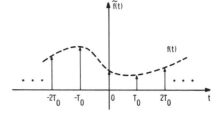

Fig. 5.2. The result of sampling the function $f(t)$ with the sampling function $s(t)$.

by the delta function simply changes the area of the delta function by the
value of $f(t)$ at the time of occurrence of the delta function. That is,

$$f(t)\,\delta(t - mT_0) = f(mT_0)\delta(t - mT_0) \qquad (5.4)$$

Therefore, a function $f(t)$ may be periodically sampled by multiplying it by
the sampling function with the result that

$$\tilde{f}(t) = f(t)s(t) = \sum_{m=-\infty}^{\infty} f(mT_0)\,\delta(t - mT_0) \qquad (5.5)$$

A typical example of $\tilde{f}(t)$ is shown in Fig. 5.2.

It is often useful to consider the frequency spectrum of a function.
From our knowledge of Fourier transformations we know that the transform of the product of two functions in direct space is a convolution in
reciprocal space. Consequently, the spectrum of $\tilde{f}(t)$ is

$$\tilde{F}(\omega) = F(\omega) * S(\omega)$$

$$= \omega_0 \sum_{m=-\infty}^{\infty} F(\omega - m\omega_0) \qquad (5.6)$$

Thus, the spectrum of the sampled function is the sum of replicas of the
spectrum of the original function centered on the points of the reciprocal
lattice as shown in Fig. 5.3 for various sampling intervals. The example
shows a spectrum $\tilde{F}(\omega)$ where the spectrum of the original function has a
cutoff at $|\omega| = W$.

It is obvious, from Fig. 5.3a, that in order to retrieve $f(t)$ from $\tilde{f}(t)$ we
need only provide an aperture or filter in reciprocal space that will exclude
all but the central spectrum. It is also obvious that this can be done only
if the repeated spectra do not overlap. Thus, we can simply state a sampling
theorem as follows: Any bandlimited function $f(t)$ can be recovered from
its periodic samples $\tilde{f}(t) = f(nT_0)$ if the sample spacing T_0, is less than or
equal to one-half the shortest wavelength in the function ($T_0 \leq \pi/W$). The

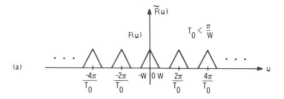

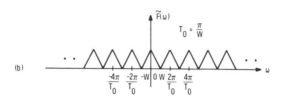

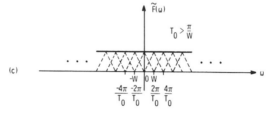

Fig. 5.3. Multiple spectra produced by the sampling
process. Aliasing occurs when the sampling interval T_0
becomes greater than π/W where W is the angular
band width of the sampled function.

statement of the sampling theorem in mathematical terms in reciprocal
space is

$$F(\omega) = G(\omega)\tilde{F}(\omega) = G(\omega)\omega_0 \sum_{m=-\infty}^{\infty} F(\omega - 2mW) \qquad (5.7)$$

where

$$G(\omega) = 1/\omega_0 \quad \text{for} \quad |\omega| \leq W$$
$$= 0 \quad \text{elsewhere}$$

The equivalent statement in direct space is

$$f(t) = g(t) * f(t) = \frac{\sin Wt}{\pi t} * \sum_{m=-\infty}^{\infty} f\left(\frac{m\pi}{W}\right)\delta\left(t - \frac{m\pi}{W}\right)$$

$$= \sum_{m=-\infty}^{\infty} f\left(\frac{m\pi}{W}\right) \frac{\sin W\left(t - \frac{m\pi}{W}\right)}{W\left(t - \frac{m\pi}{W}\right)} \qquad (5.8)$$

In each of these expressions we have assumed the optimum sampling interval of π/W. Equation (5.8) says that $f(t)$ can be reconstructed from the samples $f(m\pi/W)$ by erecting a sin x/x function of height $f(m\pi/W)$ at each sample point. The particular sin x/x function is chosen so that its zeros occur at each sampling point other than the one upon which it is centered. Figure 5.4 illustrates this statement pictorially.

We have arrived at the sampling theorem by an intuitive route. A formal mathematical proof is now given. We have the bandlimited function $f(t)$ in direct space and its transform $F(\omega)$ in reciprocal space, where

$$F(\omega) = 0 \quad \text{for} \quad |\omega| \geq W \tag{5.9}$$

Since $F(\omega)$ is confined to a finite region in reciprocal space, it can be treated as though it were periodic with period $2W$. Hence, it can be expanded in a Fourier series valid within the region $|\omega| \leq W$. Thus

$$F(\omega) = 0, \qquad |\omega| > W$$

$$= \sum_{n=-\infty}^{\infty} \alpha_n \exp\left(-i\frac{\omega n\pi}{W}\right), \qquad |\omega| \leq W \tag{5.10}$$

where

$$\alpha_n = \frac{1}{2W} \int_{-W}^{W} F(\omega) \exp\left(i\frac{\omega n\pi}{W}\right) d\omega \tag{5.11}$$

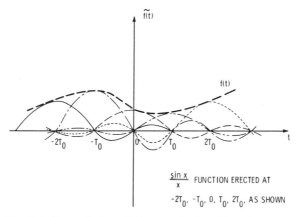

Fig. 5.4. Interpolation of the samples is achieved by erecting a (sin x)/x function at each sampling point. The height of the (sin x)/x function is the value of the function at the sampling point and the zeros occur at the sampling interval.

We know, from the Fourier transform relationship that

$$f(t) = \frac{1}{2\pi} \int_{-W}^{W} F(\omega) \exp(i\omega t) \, d\omega \tag{5.12}$$

Therefore, comparing Eqs. (5.12) and (5.11) we see that

$$\alpha_n = \frac{\pi}{W} f\left(\frac{n\pi}{W}\right) \tag{5.13}$$

Therefore, by substituting Eq. (5.13) into Eq. (5.10) we have

$$F(\omega) = \frac{\pi}{W} \sum_{n=-\infty}^{\infty} f\left(\frac{n\pi}{W}\right) \exp\left(-i\frac{\omega n\pi}{W}\right), \qquad |\omega| \leq W \tag{5.14}$$

$$= 0, \qquad |\omega| > W$$

Substituting Eq. (5.14) back into Eq. (5.12) yields

$$f(t) = \frac{1}{2W} \sum_{n=-\infty}^{\infty} f\left(\frac{n\pi}{W}\right) \int_{-W}^{W} \exp\left[i\omega\left(t - \frac{n\pi}{W}\right)\right] d\omega$$

$$= \sum_{n=-\infty}^{\infty} f\left(\frac{n\pi}{W}\right) \frac{\sin W(t - n\pi/W)}{W(t - n\pi/W)} \tag{5.15}$$

Therefore, we see that the function $f(t)$ can be reconstructed from its samples spaced at $t = n\pi/W$ by means of the simple interpolation formula of Eq. (5.15). Since Eq. (5.15) can be written in the form

$$f(t) = \left\{ \sum_{n=-\infty}^{\infty} f\left(\frac{n\pi}{W}\right) \delta(t - n\pi/W) \right\} * \left\{ \frac{\sin Wt}{Wt} \right\} \tag{5.16}$$

we recognize that $f(t)$ is recovered from its samples by convolving them with the function $(\sin Wt)/Wt$. The significance of this is that the transform of $(\sin Wt)/Wt$ is a rectangular aperture of width $2W$ centered on $\omega = 0$. The operation in reciprocal space is particularly simple since it involves a multiplication by a constant.

In summary, the sampling theorem tells us how many samples of a function we need to take in order to preserve all the information necessary to reconstruct the function. If we sample only amplitude, then the number of samples required is TW/π, where T and $2W$ are the extent of the function in direct and reciprocal space, respectively. If we consider cycles/unit, B, rather than radians/unit, W, where $W = 2\pi B$, we get the usual result of $2TB$ samples. One often speaks of this as the number of degrees of freedom in a function, or independent pieces of information.

It is not necessary that the pieces of information all be amplitude [4,5]. One could sample amplitude and the first derivative at each point. It has been shown that in this case the sample points can be twice as far apart. Indeed, if we consider Taylor's expansion theorem, we see that if amplitude and $(2TB - 1)$ derivatives at one point are sampled, we would still have our $2TB$ pieces of information.

In practice, we cannot build a sampler of infinitely small aperture as required to obtain $f(n\pi/W)$ from $f(t)$. In general, the sampler will consist of a small aperture, over which the function $f(t)$ is integrated. The value of the integral is then assigned to the height of another pulse. The new pulses, which may be of a different spacing than the original samples, then make up a new pulse train from which we wish to extract the original function. The mathematical formulation follows.

We sample the function $f(t)$ at a point $t = nT_0$ by integrating $f(t)$ over a pulse $p(t)$ centered at $t = nT_0$. That is,

$$a(nT_0) = \int_{-\infty}^{\infty} f(t)p(t - nT_0)\, dt \qquad (5.17)$$

The numbers $a(nT_0)$ are assigned as the heights of a string of pulses $h(t)$ separated by the interval T_0'. Therefore, we have a new function described by

$$z(t) = \sum_{n=-\infty}^{\infty} z(nT_0')h(t - nT_0') \qquad (5.18)$$

where

$$z(nT_0') = a(nT_0)$$

Substituting the values for $a(nT_0)$ from Eq. (5.17) we have

$$z(t) = \sum_{n=-\infty}^{\infty} \int f(t)p(t - nT_0)\, dt\, h(t - nT_0') \qquad (5.19)$$

Analogous to the process discovered in deriving the sampling theorem, we wish to extract $f(t)$ from $z(t)$ by convolving a function $g(t)$ with $z(t)$, that is

$$\tilde{f}(t) = z(t) * g(t) \qquad (5.20)$$

Taking the Fourier transform of Eq. (5.20) we have

$$\tilde{F}(\omega) = Z(\omega)G(\omega) \qquad (5.21)$$

where $G(\omega)$ is our unknown function in reciprocal space. We proceed to evaluate $Z(\omega)$ as follows:

$$Z(\omega) = \int_{-\infty}^{\infty} z(t)\,\exp(-i\omega t)\,dt$$

$$= H(\omega) \sum_{n=-\infty}^{\infty} \exp(-i\omega n T_0') \int_{-\infty}^{\infty} f(t)p(t - nT_0)\,dt \qquad (5.22)$$

We can write the integral in reciprocal space as

$$\int_{-\infty}^{\infty} f(t)p(t - nT_0)\,dt = \frac{1}{2\pi} \int_{-\infty}^{\infty} F(\mu)P^*(\mu)\,\exp(-i\mu n T_0)\,d\mu \qquad (5.23)$$

Substituting Eq. (5.23) into Eq. (5.22), we have

$$Z(\omega) = \frac{H(\omega)}{2\pi} \int_{-\infty}^{\infty} F(\mu)P^*(\mu) \sum_{n=-\infty}^{\infty} \exp[-i(\mu n T_0 + \omega n T_0')]\,d\mu \qquad (5.24)$$

Expressing T_0' as kT_0 where k is a constant, the summation can be written

$$\sum_{n=-\infty}^{\infty} \exp[-i(k\omega + \mu)nT_0]$$

which by Eq. (5.3) can be expressed as

$$\omega_0 \sum_{n=-\infty}^{\infty} \delta(k\omega + \mu - n\omega_0) \qquad (5.25)$$

where $\omega_0 = 2\pi/T_0$. When Eq. (5.25) is substituted into Eq. (5.24) $Z(\omega)$ can finally be evaluated to yield

$$Z(\omega) = \frac{H(\omega)}{2\pi}\,\omega_0 \sum_{n=-\infty}^{\infty} F(k\omega - n\omega_0)P^*(k\omega - n\omega_0) \qquad (5.26)$$

This expression shows that the spectrum of the new function $z(t)$ consists of a periodic function on the compressed reciprocal scale $k\omega$. The function is the spectrum of the original function modified by the spectrum of the sampling pulse. Riding over all the repeated spectra is that of the new pulse $h(t)$. Figure 5.5 is a diagram of the results. Formally proceeding as we did earlier, we isolate the central spectrum. This can be done if

(1) $\quad \omega_0 \geq 2W/k$

(2) $\quad G(\omega) = \dfrac{2\pi}{\omega_0 H(\omega)P^*(k\omega)}, \quad |\omega| < W/k \qquad (5.27)$

$\qquad\qquad = 0, \quad |\omega| > W/k$

For the example shown in Fig. 5.5 since $p(t)$ is even and real, $P(\omega)$ is also

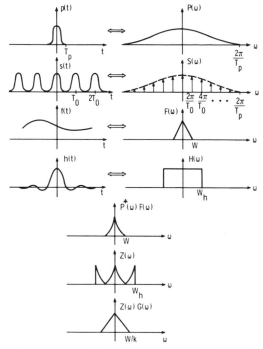

Fig. 5.5. This figure is an attempt to illustrate the general procedure for sampling and reconstructing the function $f(t)$. $p(t)$ is the sampling pulse shape, $s(t)$ the sampling function consisting of a row of pulses, $f(t)$ the function, $h(t)$ the pulse shape of the recording transducer. The spectrum of the final set of recorded samples is shown as $Z(\omega)$. The spectrum $G(\omega)$ of the interpolation function $g(t)$ then isolates the central order and corrects for the distortion introduced by $p(t)$ and $h(t)$. The scale is changed since the spacing of the recording pulses was changed.

even and real; therefore, $P^*(\omega) = P(\omega)$. The aperture weighting function, $G(\omega)$, first discards all but the central spectrum of $Z(\omega)$ and then compensates for the amplitude and phase distortions introduced by the pulses $h(t)$, $p(t)$. The final result of this operation is

$$\tilde{F}(\omega) = F(k\omega) \tag{5.28}$$

If we take the inverse transform of Eq. (5.28), we obtain

$$\tilde{f}(t) = \frac{1}{k} f\left(\frac{t}{k}\right) \tag{5.29}$$

Therefore, we have a replica of the original function on a different scale in direct space.

The above discussion may seem academic, but in fact it is exactly what we do in scanned acoustical holography. Our sampling occurs by scanning a transducer over a plane in an acoustic interference field, if we use an acoustic reference. This means that at each position, nT_0, the transducer, several wavelengths in size, integrates the intensity over its aperture, $p(t - nT_0)$. The measured value is a voltage $a(nT_0)$ which then is applied to a light source whose intensity is a function of $a(nT_0)$. The light is piped by some means to the photographic film, arriving there with its own characteristic shape, which together with the film response yields a spot of size and shape, $h(t)$. Since the light source may be moved more slowly than the transducer, we introduce the altered sample spacing T_0'. Thus, we see that far from being academic, the analysis is vital for understanding the system.

Equation (5.27) yields other important information. For $G(\omega)$ to be realizable, $H(\omega)P^*(k\omega)$ cannot be zero within the limits $|\omega| \leq W/k$, since $G(\omega)$ would have to become infinite at those points. This, then provides us with limits on $H(\omega)$ and $P(\omega)$. Since, $h(t)$, $p(t)$ are usually real and even with finite extent, $H(\omega)$ and $P(\omega)$ will have zeros. For example, a common sampler is a transducer with flat weighting. That is,

$$p(t) = K, \qquad |t| \leq T_p \tag{5.30}$$
$$= 0, \qquad |t| > T_p$$

Therefore,

$$P(\omega) = 2KT_p \frac{\sin \omega T_p}{\omega T_p} \tag{5.31}$$

The first zero of this function occurs at

$$\omega = \pi/T_p \tag{5.32}$$

Consequently, if we wish the first zero to be outside of the support W, of $F(\omega)$, we must restrict the size T_p to

$$T_p \leq \pi/W \tag{5.33}$$

Expressing W in terms of spatial wavelength, we obtain

$$2T_p \leq \lambda_0 \tag{5.34}$$

where λ_0 is the shortest spatial wavelength in the plane of sampling. Similarly, $H(\omega)$ must be nonzero within the limits $|\omega| \leq W/k$. This restricts the spot size on the film to be

$$2T_h \leq k\lambda_0 \tag{5.35}$$

5.1.2. Two-Dimensional Sampling Theory

Most of the theory developed above is directly applicable to two-dimensions if the sampling lattice is orthogonal. For complete generality, however, we need to allow periodic sampling where the lattice is oblique.

Suppose we have the two-dimensional bandlimited function $f(\mathbf{r})$ having the two-dimensional Fourier transform

$$F(\boldsymbol{\omega}) = \int_{-\infty}^{\infty} f(\mathbf{r}) \exp(-i\mathbf{r}\cdot\boldsymbol{\omega})\, d\mathbf{r}$$

$$= \int_{-\infty}^{\infty} \int_{-\infty}^{\infty} f(x, y) \exp[-i(\omega_x x + \omega_y y)]\, dx\, dy \tag{5.36}$$

where $\mathbf{r} = \mathbf{i}x + \mathbf{j}y$, $\boldsymbol{\omega} = \mathbf{i}\omega_x + \mathbf{j}\omega_y$ and $\mathbf{i}$, $\mathbf{j}$ are the unit orthogonal basis vectors. The transform $F(\boldsymbol{\omega})$ is nonzero on a finite region of the reciprocal space $\boldsymbol{\omega}$, due to the bandlimited nature of $f(\mathbf{r})$. This region, called the support of $F(\omega)$, may be a single region or it may be disconnected as shown in Fig. 5.6. We wish to sample $f(\mathbf{r})$ on a periodic lattice and determine the minimum number of lattice or sampling points required to reconstruct $f(\mathbf{r})$ from its samples, and the correct interpolation function.

We characterize the sampling lattice by means of the expression

$$\mathbf{a}_{mn} = m\mathbf{a}_1 + n\mathbf{a}_2 \tag{5.37}$$

where $n, m = 0, \pm 1, \pm 2,\ldots$ and $\mathbf{a}_1$, $\mathbf{a}_2$ are the basis vectors, not necessarily orthogonal. Figure 5.7 illustrates this expression clearly. The sample values of $f(\mathbf{r})$ are $f(\mathbf{a}_{mn})$. In complete analogy to the one-dimensional theory, we

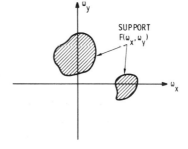

Fig. 5.6. A possible support for the spectrum of a signal.

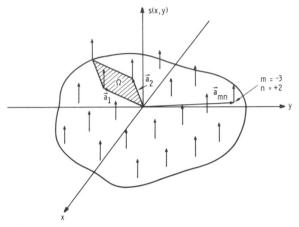

Fig. 5.7. This figure illustrates a two-dimensional oblique
sampling lattice with basis vectors $\mathbf{a}_1$, $\mathbf{a}_2$.

designate a two-dimensional sampling function

$$s(\mathbf{r}) = \sum_m \sum_n \delta(\mathbf{r} - \mathbf{a}_{mn}) \tag{5.38}$$

The Fourier transform of this function is

$$S(\boldsymbol{\omega}) = \sum_m \sum_n \exp(\boldsymbol{\omega} \cdot \mathbf{a}_{mn}) \tag{5.39}$$

$$= \frac{4\pi^2}{\Omega} \sum_m \sum_n \delta(\boldsymbol{\omega} - \mathbf{b}_{mn}) \tag{5.40}$$

where

$$\mathbf{b}_{mn} = m\mathbf{b}_1 + n\mathbf{b}_2 \tag{5.41}$$

$$\mathbf{b}_1 \cdot \mathbf{a}_2 = \mathbf{b}_2 \cdot \mathbf{a}_1 = 0 \tag{5.42}$$

$$\mathbf{b}_1 \cdot \mathbf{a}_1 = \mathbf{b}_2 \cdot \mathbf{a}_2 = 1 \tag{5.43}$$

and Ω is the area of the direct basis parallelogram. Equations (5.42) and
(5.43) describe a reciprocal lattice orthogonal to the direct lattice, having
basis vectors of length

$$b_1 = \frac{2\pi}{a_1 \sin \psi}, \qquad b_2 = \frac{2\pi}{a_2 \sin \psi} \tag{5.44}$$

where ψ is the angle between basis vectors $\mathbf{a}_1$, $\mathbf{a}_2$. Figure 5.8 illustrates the
relationships just discussed. Note that the areas of the basis parallelograms
are reciprocally related. These relationships are well known and have found

Fig. 5.8. The relationship between basis vectors in direct and reciprocal space showing that the area of the basis parallelograms are inversely related.

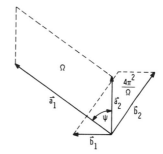

great use in crystallography [6]. We see that the sampling function is an array of impulses in both the direct and reciprocal space.

The function $f(\mathbf{r})$ is sampled by multiplying it by the sampling function to yield the new function

$$\tilde{f}(\mathbf{r}) = f(\mathbf{r})s(\mathbf{r}) - \sum_m \sum_n f(\mathbf{a}_{mn})\, \delta(\mathbf{r} \quad \mathbf{a}_{mn}) \tag{5.45}$$

The spectrum of $\tilde{f}(\mathbf{r})$ is

$$\tilde{F}(\boldsymbol{\omega}) = F(\boldsymbol{\omega}) * S(\boldsymbol{\omega})$$

$$= \frac{4\pi^2}{\Omega} \sum_m \sum_n F(\boldsymbol{\omega} - \mathbf{b}_{mn}) \tag{5.46}$$

A possible example of $\tilde{F}(\boldsymbol{\omega})$ is shown in Fig. 5.9. We see that the spectrum of the sampled function is the two-dimensional repetition of the original

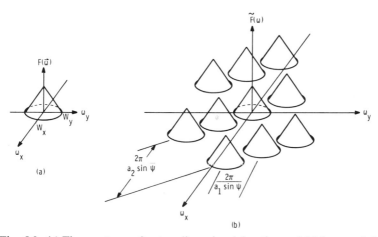

Fig. 5.9. (a) The spectrum of a two-dimensional function and (b) its sampled version when the sampling lattice is linear and periodic.

spectrum $F(\omega)$ centered on the points of the reciprocal lattice. Obviously, to recover $F(\omega)$ from $\tilde{F}(\omega)$ we need only isolate the central spectrum. This can be done by an aperture $G(\omega)$ whose support is the same as that of $F(\omega)$. This presupposes that the direct lattice points were close enough together to prevent overlap of the repeated spectrum.

It is obvious that since the spectrum may be asymmetric ($W_x \neq W_y$), we can tailor the sampling matrix for most efficient sampling. A simple example of this is shown in Fig. 5.10. For this particular spectrum, the support is rectangular. The corresponding interpolation function to use on the samples will therefore be

$$g(\mathbf{r}) = \frac{\sin W_x x}{W_x \pi_x} \frac{\sin W_y y}{W_y \pi_y} \tag{5.47}$$

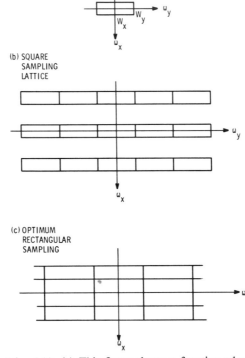

Fig. 5.10. (a) This figure shows a function whose spectrum has rectangular support, (b) the spectrum of this same function when sampled on a square lattice, and (c) when sampled on an optimum rectangular lattice.

resulting in the interpolation formula

$$f(\mathbf{r}) = g(\mathbf{r}) * f(\mathbf{r})$$

$$= \sum_m \sum_n f\left(\frac{m\pi}{W_x}, \frac{n\pi}{W_y}\right) \frac{\sin W_x\left(x - \dfrac{m\pi}{W_x}\right)}{W_x\left(x - \dfrac{m\pi}{W}\right)} \cdot \frac{\sin W_y\left(y - \dfrac{n\pi}{W_y}\right)}{W_y\left(y - \dfrac{n\pi}{W_y}\right)} \quad (5.48)$$

For the practical case of a finite sampler size and a new recording lattice, we have the result

$$G(\omega) = \frac{\Omega}{4\pi^2 H(\omega) P^*(k\omega)}, \quad |\omega| \text{ on support of } F(\omega/k)$$

$$= 0 \text{ elsewhere} \quad (5.49)$$

Information retrievable from this expression is again related to the size and shape of the sampling and recording apertures $p(\mathbf{r})$ and $h(\mathbf{r})$. From Figs. 5.9 and 5.10 it is obvious that the shape of the support of $F(\omega)$ should determine the shape of $P(\omega)$ and $H(\omega)$. A circular support requires a transducer of circular shape, since its Fourier transform will also be circular. If

$$p(\mathbf{r}) = K, \quad |\mathbf{r}| \leq r_p$$

$$= 0 \text{ elsewhere} \quad (5.50)$$

then

$$P(\omega) = 2Kr_p^2 \frac{J_1(\omega r_p)}{\omega r_p} \quad (5.51)$$

where $\omega = \sqrt{\omega_x^2 + \omega_y^2}$ and $r = \sqrt{x^2 + y^2}$. The first zero of this circular function occurs at

$$\omega r_p = 1.22\pi \quad (5.52)$$

Since we want the support of $P(\omega)$ to cover the support of $F(\omega)$ we have

$$r_p \leq \frac{1.22\pi}{W} \quad (5.53)$$

The diameter of the circular sampling transducer must therefore be

$$2r_p \leq 1.22\lambda_0 \quad (5.54)$$

The example of a rectangular support for $F(\omega)$ shown in Fig. 5.10 requires a rectangular aperture.

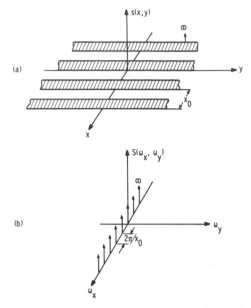

Fig. 5.11. (a) A line-scan sampling function and
(b) its spectrum.

5.1.3. Scanned Holography as a Sampling Process

As we have stated previously, most recording processes are by nature sampling processes because of the limitations of materials. Photographic film, for instance, consists of silver grains which respond more or less as a unit. Therefore, film may be thought of as a random array of detectors of finite size.

In scanned holography, the acoustical field is sampled in one-dimension and is continuously scanned in the other. Consequently, we expect to suffer the repetitive spectrum in one-dimension as discussed in the last section. Figure 5.11 pictures the sampling function in both direct and reciprocal space. Note that in reciprocal space, the sampling function consists of a line of delta functions orthogonal to the scan lines.

5.1.4. Effects of Sampling

We have not discussed previously what happens if we do not bother to eliminate all but the central spectrum of a sampled function. We will consider this problem for a perfect sampler. The spectrum of the sampled function was found to be

$$\tilde{F}(\omega) = \frac{4\pi^2}{\Omega} \sum_m \sum_n F(\omega - \mathbf{b}_{mn}) \tag{5.55}$$

$$= \frac{4\pi^2}{\Omega} \sum_m \sum_n \int f(\mathbf{r}) \exp[-i(\omega - \mathbf{b}_{mn}) \cdot \mathbf{r}] \, d\mathbf{r} \tag{5.56}$$

The equation for a unit amplitude plane wave propagating in the $\mathbf{K}$ direction is

$$B(\mathbf{R}) = \exp(-i\mathbf{K} \cdot \mathbf{R}) \tag{5.57}$$

where

$$\mathbf{K} = k_x \mathbf{i} + k_y \mathbf{j} + k_z \mathbf{k}, \quad \mathbf{R} = R_x \mathbf{i} + R_y \mathbf{j} + R_z \mathbf{k}, \quad k_z = \sqrt{1 - (k_x^2 + k_y^2)}$$

$K = 2\pi/\lambda$, and $(\mathbf{i}, \mathbf{j}, \mathbf{k})$ are the unit vectors of the x, y, z space. If we evaluate this expression in the plane $z = 0$, we have $R_z = 0$ with the result that

$$B(\mathbf{R})_{z=0} = \exp(-i\mathbf{K}' \cdot \mathbf{r}) \tag{5.58}$$

where $\mathbf{K}' = k_x \mathbf{i} + k_y \mathbf{j}$ and $\mathbf{r} = R_x \mathbf{i} + R_y \mathbf{j}$. Comparing Eq. (5.58) with Eq. (5.56), we see that we can interpret the latter as a sum of plane waves of amplitude $f(r) \, dr$ with propagation vector $\mathbf{K} = (\mathbf{K}' \cdot \mathbf{r})(\mathbf{r}/r) + k_z \mathbf{k}$, where $\mathbf{K}' = (\omega - \mathbf{b}_{mn})$ and $k_z = \sqrt{1 - (K')^2}$. Therefore, we see that the reconstruction from a sampled wavefront representing an object, will be a matrix of images of that object distributed in space. If the sampling interval is small enough, as discussed earlier, the images will be separated. If the interval is too large, the multiple images will overlap. Figures 5.12 and 5.13 illustrate these statements.

Since $\mathbf{b}_{mn}$ is periodic on an oblique lattice, $\mathbf{K}'$ will also be periodic. Therefore, the images will constitute a regular array in image space with the image lattice parallel to the reciprocal space lattice and orthogonal to the direct lattice. Figure 5.12, for example, shows this rather well. Since the direct lattice has horizontal lines, the images are arranged vertically.

A simple physical explanation is in order. As we saw in Chapter 1, a grating will diffract light in certain preferred directions dependent upon the wavelength and the grating orientation and spacing. Since the direct lattice represents a two-dimensional grating superimposed upon the hologram, we get diffraction from it. That is, the light normally diffracted into an image is further diffracted by the direct lattice into multiple images arranged orthogonally to the grating lines. So we see that, whereas the mathematics was complicated, the physics is not.

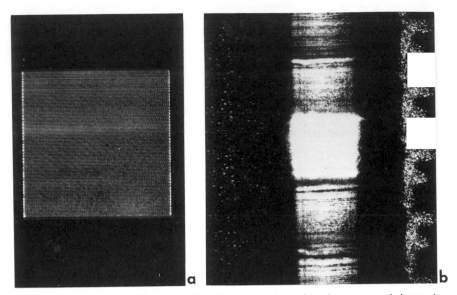

Fig. 5.12. A hologram and its image for a line-scan system with adequate sample interval.

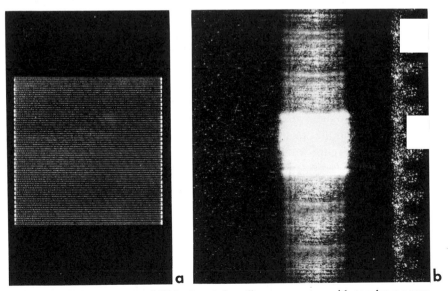

Fig. 5.13. (a) A hologram and (b) its image for a line-scan system with too large a sampling interval. Note the image aliasing.

5.2. INFORMATION CONTENT OF HOLOGRAMS

In the previous section we alluded to "bits of information" or "degrees of freedom" of a one- or two-dimensional function as being the total number of samples required to reconstruct the original bandlimited function. The first term is unacceptable since the number of samples does not reflect the meaning of information theory as developed by Shannon [1]. The second term is more meaningful for our purposes, and after a short digression, is the one we shall use here.

Information theory as developed by Shannon relates to the surprise value of receiving a "message." In terms of images, revealing the presence of an unknown flaw in a critical casting is a highly informative event with a correspondingly high measure of information.

The hologram itself, being an image encoded by Fresnel diffraction into a completely unrecognizable pattern, will always present an image unpredictable from the hologram. If one were handed a pack of holograms, one would not be able to differentiate one from another. In this sense, an ensemble of holograms represents a maximum entropy source of information. However, from the ensemble of images reconstructed from the holograms there may be no surprises at all. Hence, the classical concept of information does not really apply very well in this case.

The concept of degrees of freedom is much more applicable to our needs since it directly relates to the sophistication of equipment required to provide a satisfactory image. As was shown in our discussion of sampling, a hologram requires $4AB$ samples to be taken for all information to be recorded. In this expression, A is the aperture area, and B is the area of the support of the spatial frequency spectrum.

5.2.1. Number of Degrees of Freedom in a Scanned Receiver Hologram

Consider the arrangement shown in Fig. 5.14. From our previous work we know that the phase of the signal recorded at an arbitrary point x in the recording plane is

$$\phi(x) - k_1(r_1' - r_2') \qquad (5.59)$$

where

$$r_1' = [z_1^2 + (x - x_1)^2]^{\frac{1}{2}}$$

and

$$r_2' = [z_2^2 + (x - x_2)^2]^{\frac{1}{2}}$$

The spatial frequency at the point x is

$$\omega_x = \frac{\partial \phi(x)}{\partial x} = k_1 \left(\frac{x - x_1}{r_1'} - \frac{x - x_2}{r_2'} \right) \tag{5.60}$$

If we assume Fresnel diffraction, we can replace r_1', r_2' by r_1, r_2 to yield

$$\omega_x = mx + b \tag{5.61}$$

where

$$m = k_1 \left(\frac{1}{r_1} - \frac{1}{r_2} \right)$$

and

$$b = k_1 \left(\frac{x_2}{r_2} - \frac{x_1}{r_1} \right) = k_1(\cos \alpha_2 - \cos \alpha_1)$$

This, of course, is the equation for a straight line. If the object is other than a point, both m and b change. We can find the range of ω_x by noting that the object points must be restricted to the shaded region of Fig. 5.14, if twin image separation is to be accomplished. The slope m, is therefore constrained to

$$-\frac{k_1}{r_2} \leq m \leq k_1 \left(\frac{1}{r_{1m}} - \frac{1}{r_2} \right) \tag{5.62}$$

and the intercept to

$$0 \leq b \leq (1 + \cos \alpha_2) \tag{5.63}$$

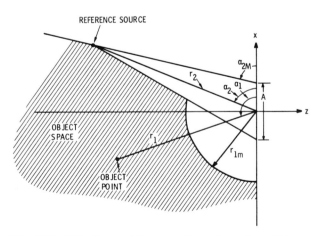

Fig. 5.14. This figure delineates the volume in which an object may lie and still be imaged by a holographic system.

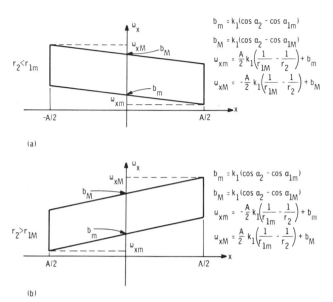

$$b_m = k_1(\cos \alpha_2 - \cos \alpha_{1m})$$

$$b_M = k_1(\cos \alpha_2 - \cos \alpha_{1M})$$

$$\omega_{xm} = \frac{A}{2} k_1 \left(\frac{1}{r_{1m}} - \frac{1}{r_2} \right) + b_m$$

$$\omega_{xM} = -\frac{A}{2} k_1 \left(\frac{1}{r_{1M}} - \frac{1}{r_2} \right) + b_M$$

(a)

$$b_m = k_1(\cos \alpha_2 - \cos \alpha_{1m})$$

$$b_M = k_1(\cos \alpha_2 - \cos \alpha_{1M})$$

$$\omega_{xm} = -\frac{A}{2} k_1 \left(\frac{1}{r_{1m}} - \frac{1}{r_2} \right) + b_m$$

$$\omega_{xM} = \frac{A}{2} k_1 \left(\frac{1}{r_{1m}} - \frac{1}{r_2} \right) + b_M$$

(b)

Fig. 5.15. This is the space–frequency diagram for (a) the reference source at a distance greater than the minimum object distance, and (b) the reference closer than the minimum object distance.

where the lower and upper extremes for the slope occur when $r_1 = \infty$, r_{1m} and for the intercept when $\alpha_1 = \alpha_2$, π, respectively, and r_{1m} is the minimum distance for which the Fresnel approximation is valid. Note also that the greatest spatial frequency W, is attained when the intercept is a maximum and the slope is maximum. Therefore, if we know the region within which the object lies, we can find W by using those values of possible r_1, α_1 which maximize m and b. For example, if the object is known to lie in the sector $\alpha_2 < \alpha_{1m} \le \alpha_1 \le \alpha_{1M}$, $r_{1m} \le r_1 \le r_{1M}$ we could draw the diagrams shown in Fig. 5.15. Note that the figures are parallelograms whose sides, parallel to the ω_x-axis, are a constant determined by the angular position and extent of the object volume. That is,

$$\Delta b = b_M - b_m = k(\cos \alpha_{1m} - \cos \alpha_{1M}) \tag{5.64}$$

independent of object range. Note also that no part of the parallelogram lies in the negative ω_x region. This, of course, is the result of using side-band holography.

A study of Fig. 5.15 reveals that if we place the reference source at the mean inverse of all possible object distances we can reduce the maximum

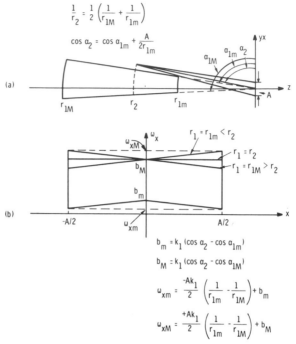

Fig. 5.16. The space–frequency diagram (b), for the geometry shown in (a). The reference source distance is at the mean inverse of the object sector distances.

slope. That is, set

$$\frac{1}{r_2} = \frac{1}{2}\left(\frac{1}{r_{1M}} + \frac{1}{r_{1m}}\right) \tag{5.65}$$

Furthermore, although Δb cannot be reduced, b_M and hence ω_{xM} can be reduced by placing the reference source near the object volume. By geometrical construction, this means that

$$\cos \alpha_2 = \cos \alpha_{1m} + \frac{A}{2r_{1m}} \tag{5.66}$$

where A is the size of the aperture in the x dimension.

If we do this for the system shown in Fig. 5.17a we might get a space-frequency diagram as shown in Fig. 5.17b. Note that if the object lies on a shell of constant radius we can make $\alpha_2 = \alpha_{1m}$, $r_2 = r_1$ with the result that the diagram becomes a rectangle with the lower side on the x-axis. This arrangement results in the lowest possible bandwidth required to record the presence of an object on the designated shell.

The spatial frequency in the ω_y-direction is less, due to the fact that the reference and object sectors are allowed to overlap in the y–z plane. Consequently, one would arrange the object sector so that the reference position lies at its mean inverse distance and at its mean cosine of the angle. That is,

$$\frac{1}{r_2} = \frac{1}{2}\left(\frac{1}{r_{1M}} + \frac{1}{r_{1m}}\right) \tag{5.67}$$

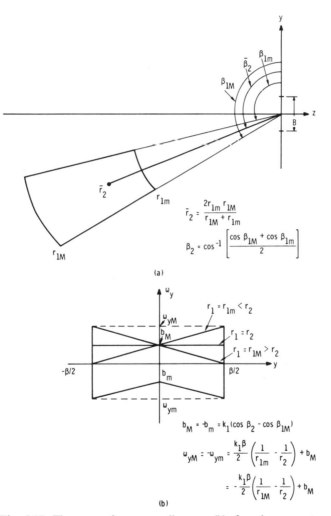

$$\bar{r}_2 = \frac{2r_{1m}\,r_{1M}}{r_{1M} + r_{1m}}$$

$$\beta_2 = \cos^{-1}\left[\frac{\cos\beta_{1M} + \cos\beta_{1m}}{2}\right]$$

(a)

$$r_1 = r_{1m} < r_2$$
$$r_1 = r_2$$
$$r_1 = r_{1M} > r_2$$

$$b_M = -b_m = k_1(\cos\beta_2 - \cos\beta_{1M})$$

$$\omega_{yM} = -\omega_{ym} = \frac{k_1\beta}{2}\left(\frac{1}{r_{1m}} - \frac{1}{r_2}\right) + b_M$$

$$= -\frac{k_1\beta}{2}\left(\frac{1}{r_{1M}} - \frac{1}{r_2}\right) + b_M$$

(b)

Fig. 5.17. The space–frequency diagram (b) for the geometry shown in (a), but with the reference source at the mean angle of the object sector. This reflects the fact that in the y dimension the reference does not need to be offset.

and

$$\cos \beta_2 = \tfrac{1}{2}(\cos \beta_{1M} + \cos \beta_{1m}) \qquad (5.68)$$

Figure 5.17 portrays both the physical arrangement and the resulting space-frequency diagram.

We can summarize the results of this section as follows. The highest spatial frequency appearing in the aperture depends upon the sector in which the object is known to lie, and the position of the reference source. We find that we can draw a space-frequency diagram which can be used to find and minimize the highest frequency. The procedure is to place the reference source at a distance such that its inverse is the mean of inverse of the extremes of the object sector. The angular position of the reference in one dimension, say x, is as close to the object volume as possible without having the hologram aperture obstructed by the object volume. In the other dimension, say y, the reference position is such that the cosine of the angle is the mean of the cosines of the extreme angles of the object volume. The resulting space-frequency diagram has a "bow-tie" shape. In the ω_x–x diagram, the "bow-tie" is entirely above the x-axis because of the reference beam offset and must never be brought below the x-axis because of twin-image overlap. In the ω_y–y diagram, the figure is centered on the origin. The narrowest part of the "bow-tie" is

$$\Delta b_x = k_1(\cos \alpha_{1m} - \cos \alpha_{1M})$$

$$= 2k_1 \sin\left(\frac{\alpha_{1M} + \alpha_{1m}}{2}\right) \sin\left(\frac{\alpha_{1M} - \alpha_{1m}}{2}\right) \qquad (5.69)$$

in the ω_x–x plane, and

$$\Delta b_y = 2k_1 \sin\left(\frac{\beta_{1M} + \beta_{1m}}{2}\right) \sin\left(\frac{\beta_{1M} - \beta_{1m}}{2}\right) \qquad (5.70)$$

in the ω_y–y plane.

Since maximum resolution is attained when the aperture is normal to the object volume, one should make $(\alpha_{1M} + \alpha_{1m})/2 = (\beta_{1M} + \beta_{1m})/2 = \pi/2$ with the result

$$\Delta b_x = 2k_1 \sin(\theta_0/2)$$
$$\Delta b_y = 2k_1 \sin(\phi_0/2) \qquad (5.71)$$

where θ_0, ϕ_0 are the angular extent of the object sector in the x–z and y–z planes, respectively. This results in a higher spatial frequency than if the hologram aperture is skewed from the perpendicular. There is a trade-off available here. If our sampling aperture is small enough to accommodate these spatial frequencies then the aperture should remain perpendicular to the object sector. If not, then one should skew the hologram aperture until the spatial frequency maximum is low enough to suit the sampling aperture. To achieve the same resolution, the hologram aperture must then be increased as the inverse of the sine of the skew angle. That is, the hologram aperture projected on a plane perpendicular to the object sector must remain the same for the same resolution. The total number of samples or degrees of freedom remains constant. This is equivalent to saying that the area of the space–frequency diagram must remain constant to maintain the same resolution. Although the maximum frequency may be reduced, the aperture must increase.

To compare various sampling schemes let us assume the case of an object lying at a distance $r_1 > r_{1m}$, somewhere in the sector $\pi/2 \pm \theta_0/2$, $\pi/2 \pm \phi_0/2$ with the reference source at $r_2 = \infty$, $\alpha_2 = \cos^{-1}[\sin(\theta_0/2) + A/2r_{1m}]$, $\beta_2 = \pi/2$. Then we have the maximum spatial frequency in the x and y dimensions of

$$W_x = 2k_1[\sin(\theta_0/2) + A/2r_{1m}] \tag{5.72}$$

and

$$W_y = k_1[\sin(\phi_0/2) + B/r_{1m}] \tag{5.73}$$

The total number of degrees of freedom then becomes

$$N = \frac{W_x W_y A}{\pi^2} = \frac{8}{\lambda_1^2} A[\sin(\theta_0/2) + A/2r_{1m}][\sin(\phi_0/2) + B/r_{1m}] \tag{5.74}$$

where A is the area of the hologram.

Finally, let us consider some numerical values. For a 10 MHz sound wave in water, a hologram 3 cm × 2 cm, an object sector 20° × 30° starting at 100 cm, the number of degrees of freedom is 11,200. The maximum spatial frequencies are $B_x = 25.2\ l/cm$ and $B_y = 18.6\ l/cm$.

5.2.2. Number of Degrees of Freedom in a Scanned Source Hologram

In the previous section we obviously had the case of a receiver sampling a stationary intensity pattern distributed in space. Consequently, the sampling lattice was a two-dimensional space lattice. When we consider the scanned source hologram, the signal is received at a fixed point in space. The signal, however, is time-varying due to the scanning source. Thus, the sampling lattice is a one-dimensional time lattice.

We were able to show that the scanning source hologram is entirely equivalent to the scanned receiver hologram with the source and receiver interchanged. Therefore, the source must occupy the same number of positions as did the receiver in the scanned receiver hologram! This certainly makes sense, since we get one reading for each position of the source. Since we need N readings or samples taken under different views of the object sector, it is obvious that we need N source positions.

When the source scanned hologram was first proposed, it was suggested that a small array of sources could be used to simulate a large number of positions by phasing techniques [8]. This was shown, on the basis of three distinct methods, to be an erroneous conclusion [9]. We appeal to an optical analogy for proof.

In Chapter 1 we discussed the zone plate and showed that it has focusing properties. If we make up an aperture containing a zone plate and illuminate it with a plane coherent source, we will obtain a wave diverging from an apparent point source. If we wish to move the apparent source we translate the zone plate parallel to itself. A large aperture is required to be able to translate the simulated source the desired amount.

The array analog to the zone plate is to produce a phase distribution as shown in Fig. 5.18a. From sampling theory we know that we can adequately reproduce this function by sampling each cycle twice. Therefore, if we provide an array of sources as shown in Fig. 5.18b with phases indicated by the direction of the arrows, we could expect to obtain a spherical wave front, plus repetition thereof at other angles. We stress that this phase distribution is the only one yielding a wave that is not just locally spherical.

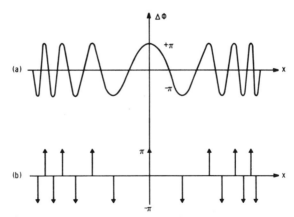

Fig. 5.18. The phase diagram (a) of a Fresnel zone pattern and its binary representation (b).

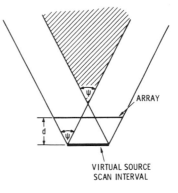

Fig. 5.19. Representation of an array used to simulate a scanning point source. The shaded region is the volume in which the simulation is good.

If we wish the array to scan the virtual point, it must either be physically moved or the phase must be redistributed. Thus, the array must be made larger to accommodate the shifted zone plate and the array elements must be closer together because the elements at the center were spaced to adequately sample the center of the zone plate and not the edges.

Thus, when all is said and done, we must have array elements spaced at twice the highest spatial frequency required for the phase distribution if we want to be able to scan electronically. From Fig. 5.19 we see that the array must be one zone plate larger than the simulated scan aperture. The sector in which a scanning point source is simulated is shown by the shaded region. For a fixed sector size, ψ, the size of the zone plate is determined by the distance, d, of the virtual source from the array. As we decrease d, we see that the array size approaches that of the virtual array size. Since the number N, of required virtual source positions depends upon the sector size ψ, and the number of elements in the zone plate depends on the same quantity, it must be true that the number of elements in the array always exceeds N except for the case of $d = 0$. Therefore, if we wish to use an electronically scanned array, the most economical in terms of number of array elements is one with N elements switched on and off in sequence. Any other array will require more than N elements. The only advantage for this type of array might be the increased power in the beam from using all elements together. The phasing problems are, however, highly complex.

The temptation to consider a simple array for the simulation of a scanned point source is great but we must realize that it must look like the same source from every point in the sector. Two elements, for instance, can be phased so that they simulate a scanning point source, but for one point in space only. In general, N elements in an array can be made to simulate a moving point source for no more than $(N - 1)$ points in space.

This fact has been proven by theoretical analysis and by computer studies at Battelle-Northwest [10]. An experiment using four sources physically scanned in one dimension and electronically scanned in the other produced the hologram shown in Fig. 5.20. The object was a specular reflector and appeared as a point, and the electronic phasing was designed for that one point only.

5.2.3. Number of Degrees of Freedom in a Scanned Source and Receiver Hologram

We consider here the case of coincident source and receiver only, on the grounds that this is the optimum system with respect to resolution. For this particular case we have the phase distribution

$$\phi(x) = k_1(2r_1' - r_2') \tag{5.75}$$

The spatial frequency becomes

$$\omega_x = \frac{\partial \phi(x)}{\partial x} \cong mx + b \tag{5.76}$$

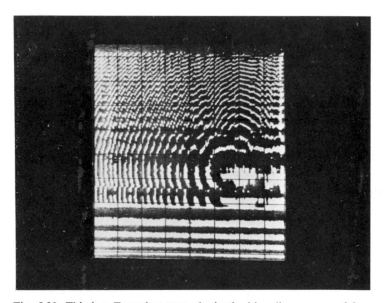

Fig. 5.20. This is a Fresnel pattern obtained with a linear array of four sources phase programmed to simulate a scanning point in the horizontal dimension. The array was towed in the vertical dimension. The object was a 55-gal oil drum and the acoustical frequency was 50 kHz. The range to the object was 30 ft.

where

$$m = k_1\left(\frac{2}{r_1} - \frac{1}{r_2}\right)$$

and

$$b = k_1(\cos \alpha_2 - 2 \cos \alpha_1)$$

Note in this case, that in order for the space–frequency diagram to remain in the upper half-plane we must fulfill the inequality

$$\cos \alpha_2 \geq 2 \cos \alpha_1 \tag{5.77}$$

This is to be expected, since the increased resolution obtained by this system must result from higher spatial frequencies and this in turn requires more offset between the reference and object beams.

The discussion of Section 5.2.1 is valid for this case, with the above limitation on object sector location. If we formally repeat the various equations, we obtain the following:

$$b_m = k_1(\cos \alpha_2 - 2 \cos \alpha_{1m}) \tag{5.78}$$

$$b_M = k_1(\cos \alpha_2 - 2 \cos \alpha_{1M}) \tag{5.79}$$

$$\omega_{xm} = \frac{Ak_1}{2}\left(\frac{2}{r_{1m}} - \frac{1}{r_2}\right) + b_m, \qquad r_2 < r_{1m}/2 \tag{5.80}$$

$$= -\frac{Ak_1}{2}\left(\frac{2}{r_{1m}} - \frac{1}{r_2}\right) + b_m, \qquad r_2 > r_{1m}/2 \tag{5.81}$$

$$\omega_{xM} = -\frac{Ak_1}{2}\left(\frac{2}{r_{1m}} - \frac{1}{r_2}\right) + b_M, \qquad r_2 < r_{1M}/2 \tag{5.82}$$

$$= \frac{Ak_1}{2}\left(\frac{2}{r_{1m}} - \frac{1}{r_2}\right) + b_M, \qquad r_2 > r_{1M}/2 \tag{5.83}$$

$$\Delta b_x = 2k(\cos \alpha_{1m} - \cos \alpha_{1M}) \tag{5.84}$$

When we optimize the reference position by the equations

$$\frac{1}{r_2} = \frac{1}{r_{1M}} + \frac{1}{r_{1m}} \tag{5.85}$$

$$\cos \alpha_2 = \cos \alpha_{1m} + A/2r_{1m}, \qquad A/2r_{1m} \geq \cos \alpha_{1m} \tag{5.86}$$

$$= 2\cos \alpha_{1m}, \qquad A/2r_{1m} < \cos \alpha_{1m}$$

we have the space-frequency diagram shown in Fig. 5.21. The expression

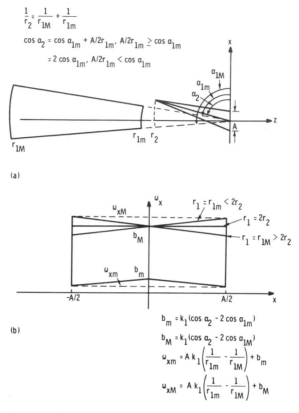

(a)

(b)

Fig. 5.21. The space–frequency diagram (b), for the geometry shown in (a). This geometry is optimum when source and receiver are scanned simultaneously. Note that for this case the reference is always closer to the hologram than the minimum object distance.

for angular position is two-valued because of the inequality in Eq. (5.77). Note also that $r_2 \leq r_{1m}$ with equality when $r_{1M} = \infty$. From the equations we see that the spatial frequency bandwidth will be on the order of twice as large as that for the scanned receiver hologram.

If we run through the example used in Section 5.2.1, we arrive at the expressions

$$W_x = 3k_1[\sin(\theta_0/2) + A/2r_{1m}], \qquad A/2r_{1m} \geq \sin(\theta_0/2)$$
$$= 4k_1[\sin(\theta_0/2) + A/4r_{1m}], \qquad A/2r_{1m} < \sin(\theta_0/2) \qquad (5.87)$$

$$W_y = 2k_1[\sin(\phi_0/2) + B/2r_{1m}] \qquad (5.88)$$

Using the numerical values from our previous example, we find that $A/2r_{1m} < \sin(\theta_0/2)$, so we must use the corresponding expression for W_x. Then we obtain a total number of samples $N = 41,800$ with maximum spatial frequencies of $W_x = 48.5\ l/\text{cm}$ and $W_y = 35.9\ l/\text{cm}$. Therefore, if we wish to use simultaneous source and receiver scanning, we must be prepared to use a smaller mesh sampling lattice. Figure 5.22 is a laboratory demonstration of this effect.

5.3. SPECIAL SAMPLING SCHEMES

In the sampling theory considered up to now, the sampling lattice has been periodic. The direct result of this assumption is that the Fourier transform of the sampled function is also periodic on the reciprocal lattice. In holography this means that multiple images are formed on yet another periodic lattice. If the direct lattice becomes too coarse, the reciprocal lattice becomes too fine with the result that the multiple images start to overlap.

There is nothing sacred about the periodic lattice other than the ease with which it can be analyzed. Similarly, in two dimensions, it is not required that the lattice be constructed of straight lines. Interestingly enough, we have been unable to discover in the literature, any examples of other than parallelogram lattices except in array theory. There are distinct advantages to avoiding the periodic parallelogram sampling lattices. We will attempt to show why this is so.

As a first step we consider rotationally symmetric sampling. It has been shown that the Fourier transform of a rotationally symmetric function is related to the Hankel transform by the relation

$$F(\omega_x, \omega_y) = 2\pi F_h(\sqrt{\omega_x^2 + \omega_y^2}) \tag{5.89}$$

where

$$F_h(\omega) = \int_0^\infty rf(r)J_0(\omega r)\, dr \tag{5.90}$$

is called the Hankel transform of zero order, $f(r)$ is the rotationally symmetric function, $r = \sqrt{x^2 + y^2}$, and $\omega = \sqrt{\omega_x^2 + \omega_y^2}$ [11]. The inverse Hankel transform is

$$f(r) = \int_0^\infty \omega F_h(\omega)J_0(\omega r)\, d\omega \tag{5.91}$$

In developing the one-dimensional sampling theory we found that it depended upon finding an interpolation function of such a form that its zeros

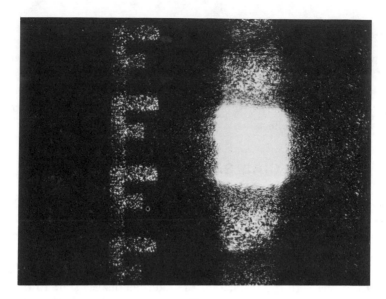

a

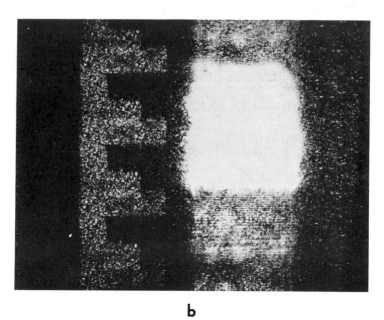

b

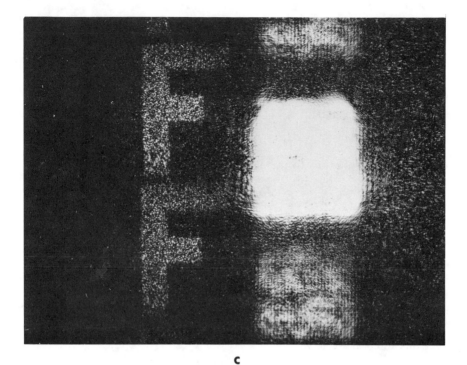

c

Fig. 5.22. These photographs illustrate the increased scan density required for the source–receiver scan hologram. The upper photograph (a) is the image from a receiver scan hologram with just adequate line density, the center (b) from a source–receiver scan hologram with the same line density, and the lower (c) from a source–receiver scan hologram with doubled line density.

coincided with all sampling points other than the one upon which it was centered. In the rectangular system of coordinates, the natural function turned out to be $(\sin \theta)/\theta$, whose zeros are periodically spaced at intervals of $\theta = \pi$.

5.3.1. A Rotationally Symmetric Sampling Theorem

For the rotationally symmetric function, cylindrical coordinates are appropriate, resulting in the Bessel functions as the natural interpolation functions. The zeros of the Bessel function, however, are not equidistant. Consequently, the sampling points will not be periodic.

The discussion to this point has been highly intuitive. We now proceed to verify these ideas mathematically. It is well known that a function can be expanded in the Fourier–Bessel series (see Ref. 12, p. 158)

$$f(x) = \sum_{m=1}^{\infty} a_m J_0(\alpha_m x) \tag{5.92}$$

over the interval $0 < x < a$ where

$$a_m = \frac{2}{a J_1^2(\alpha_m x)} \int_0^a x f(x) J_0(\alpha_m x)\, dx \tag{5.93}$$

and α_m are the solutions to the equation

$$J_0(\alpha_m a) = 0 \tag{5.94}$$

The Hankel transform of Eq. (5.92) is

$$F_h(\omega) = \sum_{m=1}^{\infty} \frac{a_m}{\alpha_m} \delta(\omega - \alpha_m) \tag{5.95}$$

as can easily be verified by performing the inverse Hankel transform. We note that the integral in the expression for a_m is the Hankel transform of $f(x)$ in the interval $0 < x < a$ with the argument a_m. If $f(x)$ exists only over this interval we have

$$a_m = \frac{2 F_h(\alpha_m)}{a^2 J_1^2(\alpha_m a)} \tag{5.96}$$

We can now apply these concepts to a sampling theorem. We assume that the function $f(x, y)$ has a spectrum $F(\omega_x, \omega_y)$ finite within or on a circular boundary $\omega_x^2 + \omega_y^2 = W^2$, and zero elsewhere. We further assume that the function is sampled by a linear array of point receivers rotated about one end. The receivers are spaced according to the zeros of the zeroth order Bessel function.

The sampling function is expressed as

$$s(x, y) = s(\sqrt{x^2 + y^2}) = s(r) = \sum_{m=0}^{\infty} \delta(r - \alpha_m) \tag{5.97}$$

where $J_0(\alpha_m W) = 0$.

The Fourier transform of the sampling function is

$$S(\omega_x, \omega_y) = 2\pi S_h(\sqrt{\omega_x^2 + \omega_y^2}) = 2\pi S_h(\omega) = 2\pi \sum_{m=0}^{\infty} \alpha_m J_0(\alpha_m \omega) \tag{5.98}$$

We next expand $F_h(\omega)$ in the Fourier–Bessel series

$$F_h(\omega) = \frac{2}{W^2} \sum_{m=1}^{\infty} \frac{f(\alpha_m)}{J_1^2(\alpha_m W)} J_0(\alpha_m \omega), \qquad |\omega| \le W \tag{5.99}$$

The inverse Hankel transform of Eq. (5.99) yields

$$f(r) = \int_0^\infty \omega F_h(\omega) J_0(\omega r) \, d\omega$$

$$= \frac{2}{W^2} \sum_{m=1}^\infty \frac{f(\alpha_m)}{J_1^2(\alpha_m W)} \int_0^W J_0(\alpha_m \omega) J_0(\omega t) \, d\omega \qquad (5.100)$$

From tables of Hankel transforms we find that the integral becomes

$$\alpha_m W \left[\frac{J_1(\alpha_m W) J_0(Wr) - r J_0(\alpha_m W) J_1(Wr)}{\alpha_m^2 - r^2} \right] \qquad (5.101)$$

But $J_0(\alpha_m W) = 0$ by definition, with the result that

$$f(r) = \frac{2}{W} \sum_{m=1}^\infty \frac{\alpha_m f(\alpha_m) J_0(Wr)}{J_1(\alpha_m W)(\alpha_m^2 - r^2)} \qquad (5.102)$$

This equation shows that a two-dimensional, band-limited function can be recovered from its samples taken at radial intervals corresponding to the zeros of the zeroth order Bessel function by means of an appropriate interpolation function.

5.3.2. Circular Scanning

In actual practice it is inconvenient to build a linear array with the receivers spaced unequally as required by the above theory. It is much more likely that they will be equally spaced, with the result that the sampling function becomes an array of concentric cylinders shown in Fig. 5.23a described by the expression

$$s(r) = \sum_{m=0}^\infty \delta(r - m r_o) \qquad (5.103)$$

By the Poisson summation formula for circularly symmetric functions we have

$$\sum_{m=0}^\infty f(r - m r_o) = \frac{1}{r_o} \sum_{m=0}^\infty F_h(m \omega_o) J_0(m \omega_o r)$$

Hence,

$$s(r) = \sum_{m=0}^\infty \delta(r - m r_o) = \frac{1}{r_o} \sum_{m=0}^\infty J_0\left(\frac{2\pi m r}{r_o} \right) \qquad (5.104)$$

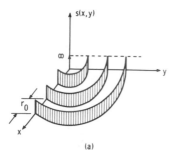

(a)

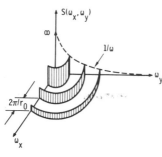

(b)

Fig. 5.23. (a) The circular symmetric sampling
function and (b) its transform.

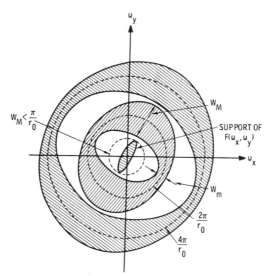

Fig. 5.24. Support of the spectrum of the function
comprised of $f(x, y)$ sampled with a circularly symmetric sampling function. Note that the spectrum of
$f(x, y)$ appears only once.

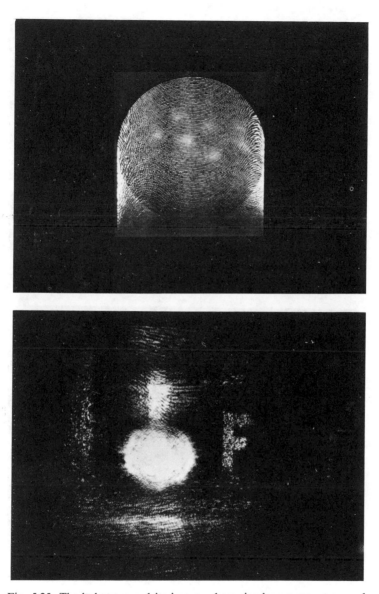

Fig. 5.25. The hologram and its image, where circular arc scan was used with adequate line density. Note that no multiple images exist, but an arc of light surrounds the image.

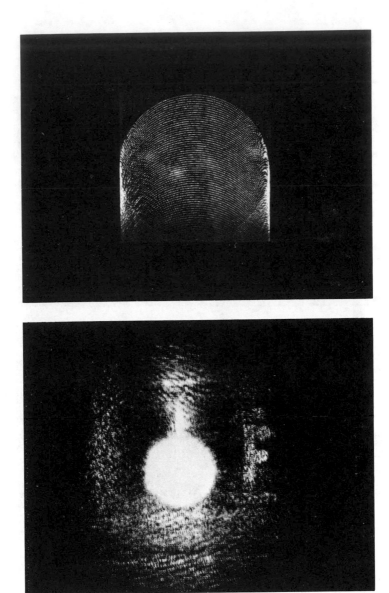

Fig. 5.26. The hologram and its image, where circular arc scan was used with inadequate line density. No aliasing occurs; rather a slight rise in background noise occurs due to the overlap of the first arc with the image.

The Fourier transform for this sampling function is

$$s(\omega) = 2\pi s_h(\omega) = \frac{2\pi}{r_0} \sum_{m=0}^{\infty} \int_0^{\infty} r J_0 \left(\frac{2\pi m r}{r_0} \right) J_0(\omega r) \, dr \qquad (5.105)$$

$$= \sum_{m=0}^{\infty} \frac{1}{m} \delta(\omega - m\omega_0)$$

where $\omega_0 = 2\pi/r_0$. Consequently, the sampling function in reciprocal space is again a set of concentric cylinders. Now, however, the heights of the cylinders decrease with radius as shown in Fig. 5.23b.

The support of the spectrum of the sampled function is illustrated in Fig. 5.24, where the original spectrum is assumed to have elliptical support. Remember that a cross section would show the maxima of each ring reduced as $1/\omega$. In the case shown, the sampling interval has been small enough for perfect separation of the central spectrum from the rings. Thus, all that is required (for the recovery of the function) is a circularly symmetric aperture in the transform plane. Note, however, that undersampling does not penalize the system nearly as much as for rectilinear scanning because there are no repetitions of the spectrum. The higher orders are smeared out over an annulus.

For example, if the scanned hologram, whose spectrum is shown in Fig. 5.24, were reconstructed we would not see multiple images. Only the central order in the spectrum focuses to an image. The ring orders focus in the radial direction, but are smeared out in the azimuthal direction resulting in a faint ring of light. If the hologram is undersampled, this ring of light moves closer to the image and finally overlaps it. However, unless gross undersampling is done so that the second and third rings overlap the image, very little image deterioration occurs. Furthermore, the deterioration is a signal-to-noise ratio (SNR) type rather than an interference or multiple image type.

A hybrid scanning system combining the simplicity of a single receiver scan, with the rapidity of a circular array scan is shown in Fig. 5.25. In this case, a single receiver is mounted on the edge of a disk. A plane is scanned by rotating and translating the disk, thereby obtaining a scan pattern of translated circular arcs. This kind of hologram retains the advantages of the circular scan holograms, including only one image. In addition it has the advantage of continuous control on the line density, merely by changing the velocity of translation. Examples of reconstructions from such holograms are shown in Figs. 5.25 and 5.26.

REFERENCES

1. C. E. Shannon, A mathematical theory of communication, *Bell System Tech. J.* **27**:379, 623 (1948).
2. D. Gabor, Light and information, in *Progress in Optics*, *Vol.* 1, E. Wolf (ed.), North-Holland Pub. Co., Amsterdam, p. 111 (1964).
3. H. J. Landau, Sampling, data transmission, and the Nyquist rate, *Proc. I.E.E.E.* **55**(10):1701 (1967).
4. D. P. Peterson and D. Middleton, Sampling and reconstruction of wavenumber-limited functions in N-dimensional Euclidean spaces, *Inf. and Cont.* **5**:279 (1962).
5. D. A. Linden, A discussion of sampling theorems, *Proc. I.R.E.* **47**:1219 (1959).
6. L. Brillouin, *Wave Propagation in Periodic Structures*, 2nd Ed., Dover Publications, New York (1953).
7. J. W. Goodman, *Introduction to Fourier Optics*, McGraw Hill, p. 48 (1968).
8. A. F. Metherell and S. Spinak, Acoustical holography of nonexistent wavefronts detected at a single point in space, *Appl. Phys. Lett.* **13**:22 (1968).
9. B. P. Hildebrand, Information content of holograms and its relationship to scanning, *Internal Report No. SSP*-67-4, Battelle-Northwest, Richland, Washington, August 1967.
10. S. C. Keeton, Wavelength scaling problems in holography, *Internal Report No. SSP*-67-2, Battelle-Northwest, Richland, Washington, August, 1967.
11. A. Papoulis, *Systems and Transforms with Applications in Optics*, McGraw-Hill, New York, p. 141 (1968).

Liquid-Surface Holography

6.1. INTRODUCTION

This chapter discusses the interaction of sound (acoustical energy) and light at a liquid surface. The specific interaction involves two beams of sound, one of which should have a wave front of simple geometry so that the wave front can easily be simulated in a beam of light. Usually this means that the wave front should be either spherical or plane. This sound beam will be referred to as the reference beam. The other beam, referred to as the object beam, must be essentially coherent* with the reference beam and should insonify the object uniformly but may have any wave front geometry.

The object to be examined will be placed in the object beam. Most objects will modify both the phase and the amplitude of the object beam in a complex manner. To simplify thinking about the imaging process, it is necessary to discuss "small" regions within the object over which the object beam is uniformly modified in phase and amplitude. These "small" regions are small with respect to the object, yet large with respect to the wavelength of the sound being used for imaging. Having treated one such small region, it is possible to consider the object as made up of a collection of such regions.

Much of the analysis presented here treats the imaging process as though it were fully continuous whereas, in fact, most imaging is carried out using pulsed sound and light with parameters such as repetition rate, pulse length,

---

* The rapid response of the liquid surface permits a few percent difference in frequency between the object and reference beams.

and the relative delays between the sound- and light-pulses being adjusted for optimum performance. Typically the wave packet of the sound pulse carries 100 cycles and the duration of the pulse of light is 10 to 80 μsec.

Figure 6.1 illustrates the typical liquid-surface acoustical holography imaging system. The system includes an object beam transducer that insonifies the object. The object scatters, diffracts, absorbs energy, and modifies the phase distribution in the wavefront. A pair of acoustic lenses are used to image the sound distribution at the object into the plane of the

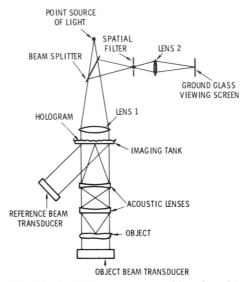

Fig. 6.1. An ideal arrangement of imaging ultrasound by use of liquid surface acoustical holography techniques employs a pair of acoustic lenses to form an image in the plane of the liquid surface hologram. At the hologram surface scattered and unscattered energy transmitted by the object is brought into interference with a plane wave reference beam. The interference pattern so formed impresses itself upon the liquid surface in the form of variations in elevation. The liquid surface, thus deformed, serves as a complex grating which diffracts light. All the light reflected and diffracted by the liquid surface is rejected by the spatial filter except for light diffracted into the first order. Lens 2 is used to image the hologram surface on the ground-glass viewing screen in first-order diffracted light.

hologram. This sound field mixes with the sound field of the reference beam transducer to form an interference pattern at the liquid–air interface in the imaging tank. The interference pattern is imprinted in the liquid surface as variations in elevation.

An imaging tank (often called a minitank) is used to isolate the hologram from liquid-surface disturbances in the main (object) tank. Use of a minitank also allows the use of different liquids for hologram formation and for object immersion. Thus, liquids can be chosen which are optimum for each of these purposes.

The hologram surface imposes phase variations in the reflected beam of light which cause diffraction of the light. First-order diffracted light is passed at the spatial filter which blocks all other light reflected from the hologram surface.

A lens (Lens 1) is used to image the point source of light at the plane of the spatial filter after reflection from the hologram surface. A second lens (Lens 2) images the hologram surface on the ground glass viewing screen. Since the hologram is a focused-image type, the ground-glass screen displays the desired image of the object. Although the hologram modifies only the phase of the light, there is amplitude imaging because the amount of light diffracted into the first order varies as the amplitude of the interference pattern on the liquid surface which, in turn, varies as the amplitude of the wave in the object beam.

6.2. DESCRIPTION OF THE ACOUSTICAL FIELD

Analytical expressions developed in this section neglect edge effects at the boundaries between domains of uniform intensity. Furthermore, the two acoustical beams are assumed to be incident at equal but opposite angles, Θ, to the normal to the liquid surface as shown in Fig. 6.2. The reference beam may be characterized by the expression

$$U_r(x, y, z) = P_r \exp[i(\eta y + \zeta z)] \tag{6.1}$$

where U_r is the amplitude of the acoustical wave at the point (x, y, z) and P_r is the maximum amplitude. Furthermore

$$\eta = (2\pi/\Lambda) \sin \Theta \tag{6.2}$$

and

$$\zeta = (2\pi/\Lambda) \cos \Theta \tag{6.3}$$

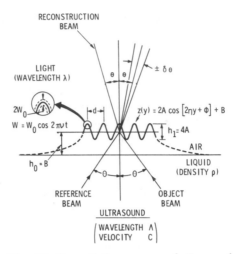

Fig. 6.2. The radiation pressure of ultrasound upon the liquid surface produces an overall bulge B and a ripple of amplitude $2A$. Relative dimensions are distorted in this figure in that the ripple amplitude A is normally orders of magnitude less than the crest-to-crest ripple spacing d. The amplitude W_0 of oscillation of the liquid surface at the frequency of the ultrasonic wave is even smaller, being normally orders of magnitude less than the ripple amplitude A.

where $\varLambda$ is the wavelength of the acoustic wave. In the reference wave the amplitude P_r is considered a constant and there is, therefore, no variation of U_r with respect to x.

Before interaction with the object, the object wave may be considered to have the same general form as the reference wave; i.e., the object wave before interaction with the object may be characterized by the expression

$$U_o(x, y, z) = P_o \exp[-i(\eta y - \zeta z)] \qquad (6.4)$$

The interaction of the acoustic wave described by Eq. (6.4) with the object introduces variations in the pressure amplitude P_o and the phase ϕ which, as was noted in Section 6.1, are constant over "small" regions in the object. The hologram may also be considered to be made up of "small" areas of uniform characteristics having centroid coordinates $(X_i Y_i)$. Thus at the liquid surface where $z = 0$, the object wave may be characterized by

$$U_o(X_i, Y_i, y) = P_o(X_i, Y_i) \exp\{-i[\eta y + \phi(X_i, Y_i)]\} \qquad (6.5)$$

and the reference wave by

$$U_r(y) = P_r \exp(i\eta y) \qquad (6.6)$$

The interference of the two acoustical beams defined by Eqs. (6.5) and (6.6) produces an intensity distribution given by

$$I(X_i, Y_i, y) = |U_o + U_r|^2/2\varrho c \qquad (6.7)$$

where ϱ is the density of the liquid and c is the velocity of sound in the liquid. Reflection of the sound at the liquid surface produces a radiation pressure Π given by [see Eq. (3.97)]

$$\Pi(X_i, Y_i, y) = 2I(X_i, Y_i, y)/c \qquad (6.8)$$

This pressure is opposed by pressures due to gravity and surface tension. Let these two pressures be denoted by Π_g and Π_t, respectively. That due to gravity may be calculated simply as the weight of a column of liquid of unit cross-sectional area and height z, i.e.,

$$\Pi_g = \varrho g z \qquad (6.9)$$

where g is the acceleration of gravity.

The pressure due to surface tension, Π_t, will be described with reference to Fig. 6.3. Consider a column of liquid of height z and cross section $dxdy$. The upward force, T_z, exerted on the column by surface tension, is given by

$$T_z = (\gamma \sin \alpha_2 - \gamma \sin \alpha_1)\, dx \qquad (6.10)$$

where γ is the surface tension in force per unit length (for water $\gamma \cong 72$ dynes/cm). When α_1 and α_2 are sufficiently small, as they will be in this

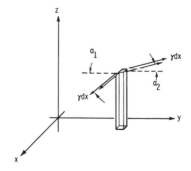

Fig. 6.3. Surface tension or free surface energy γ produces a vertical force T_z on the column of liquid of height z and cross section $dxdy$. An upward force $\gamma \sin \alpha_2\, dx$ is opposed by a downward force $\gamma \sin \alpha_1\, dx$. The net result is a force T_z which is the difference between these two forces.

instance,

$$\sin \alpha_1 = \tan \alpha_1 = \frac{\partial z}{\partial y} \tag{6.11}$$

and

$$\sin \alpha_2 = \sin \alpha_1 + \frac{\partial \sin \alpha_1}{\partial y} \, dy \tag{6.12}$$

so that the force due to surface tension is

$$T_z = \gamma \, \frac{\partial^2 z}{\partial y^2} \, dxdy \tag{6.13}$$

The equivalent pressure due to surface tension is

$$T_z/dxdy = \gamma \, \frac{\partial^2 z}{\partial y^2} \tag{6.14}$$

On the basis of Eqs. (6.9) and (6.14), the local pressure balance equation may be written

$$\Pi(X_i, Y_i, y) = \varrho g z - \gamma \, \frac{\partial^2 z}{\partial y^2} \tag{6.15}$$

in which the elevation z above the quiescent level $z = 0$ of the liquid surface is a function of the coordinates X_i, Y_i and y.

Referring again to Eq. (6.7) note that

$$| U_o + U_r |^2 = (U_o + U_r)(U_o{}^* + U_r{}^*)$$
$$= U_o U_o{}^* + U_o U_r{}^* + U_r U_o{}^* + U_r U_r{}^*$$
$$= [P_o(X_i, Y_i)]^2 + P_r{}^2 + 2P_o(X_i, Y_i)P_r \cos[2\eta y + \phi(X_i, Y_i)] \tag{6.16}$$

On the basis of Eqs. (6.7), (6.8), and (6.16)

$$\Pi(X_i, Y_i, y) = \frac{2P_r}{\varrho c^2} \, P_o(X_i, Y_i) \cos[2\eta y + \phi(X_i, Y_i)]$$
$$+ \frac{P_r{}^2 + [P_o(X_i, Y_i)]^2}{\varrho c^2} \tag{6.17}$$

Assuming a stationary solution for z of the form,

$$z(X_i, Y_i, y) = 2A(X_i, Y_i) \cos[2\eta y + \phi(X_i, Y_i)] + B(X_i, Y_i) \tag{6.18}$$

and applying Eq. (6.15) yields

$$A(X_i, Y_i) = \frac{P_r P_o(X_i, Y_i)}{\varrho c^2 [\varrho g + 4\gamma \eta^2]} \tag{6.19}$$

and

$$B(X_i, Y_i) = \frac{P_r^2 + [P_o(X_i, Y_i)]^2}{\varrho^2 c^2 g} \tag{6.20}$$

The relative magnitudes of the terms in the denominator of Eq. (6.19) are such that the equation is normally well approximated by

$$A(X_i, Y_i) = \frac{P_r}{4\gamma\varrho c^2\eta^2} P_o(X_i, Y_i) \tag{6.21}$$

Figure 6.2 is a schematic representation of the liquid surface in the region of the hologram showing the parameters associated with the description of the hologram. The magnitude of z relative to the other parameters is, of necessity, greatly exaggerated. The principal ray in each of the two acoustical beams is shown. Consistent with the analysis of this chapter, the two beams are incident upon the liquid surface at equal but opposite angles Θ.

Vibration of the liquid surface at the frequency ν of the acoustical wave occurs at an amplitude W which is orders of magnitude smaller than the amplitude of the ripple pattern, which in turn is orders of magnitude less than the overall liquid elevation B.

To demonstrate the relative magnitudes of the features shown in Fig. 6.2, consider the case of a point object (source) at infinity radiating at the same intensity as a similar point source at infinity which is used to generate the reference beam. Under the conditions described, the pressure amplitudes P_r and P_o are equal so it is useful to define an average intensity I_a such that

$$2\varrho c I_a = P_o^2 = P_r^2 = P_o P_r \tag{6.22}$$

For water

$$\varrho g = 980 \text{ g cm}^{-2} \text{ sec}^{-2} \tag{6.23}$$

whereas

$$4\gamma\eta^2 = 4\gamma\left(\frac{2\pi}{\Lambda} \sin \theta\right)^2 \tag{6.24}$$

so that if $\theta = 30°$, $\gamma = 73$ dyne cm^{-1}, $c = 1.5 \times 10^5$ cm sec^{-1}, and $\nu = 10^6$ sec^{-1}

$$4\gamma\eta^2 = 1.2 \times 10^5 \text{ g cm}^{-2} \text{ sec}^{-2} \tag{6.25}$$

Thus, of the two terms in the denominator of Eq. (6.19), $4\gamma\eta^2$ is by far the larger and the relative magnitudes are such that the approximation used in Eq. (6.20) introduces only 1% error.

Under the conditions set by Eq. (6.22),

$$A(X_i, Y_i) = \frac{2\varrho c I_a}{4\gamma \varrho c^2 \eta^2} = \frac{2I_a}{4\gamma c \eta^2} \qquad (6.26)$$

and

$$B(X_i, Y_i) = \frac{4\varrho c I_a}{g\varrho^2 c^2} = \frac{4I_a}{\varrho g c} \qquad (6.27)$$

so that

$$\frac{B}{2A} = \frac{4\gamma \eta^2}{\varrho g} \simeq 120 \qquad (6.28)$$

showing that the features of the hologram are more than two orders of magnitude less in amplitude than the overall liquid elevation B. This feature of the liquid-surface hologram precludes imaging any but the simplest objects outside the hologram plane and dictates operation in the focused image mode for high-quality imaging unless other techniques are used to reduce the magnitude of the ratio in Eq. (6.28).

6.3. ACOUSTICAL TRANSFER FUNCTIONS FOR A CONTINUOUS WAVE HOLOGRAM

A transfer function relating the hologram ripple amplitude $A(X_i, Y_i)$ to the object beam pressure amplitude $P_o(X_i, Y_i)$ is given by

$$\frac{A}{P_o} = \frac{P_r}{\varrho c^2} \frac{1}{\varrho g + 4\gamma \eta^2} \qquad (6.29)$$

Expressed in dimensionless form, Eq. (6.29) becomes

$$G_p = \frac{\varrho^2 c^2 g}{P_r} \frac{A}{P_o} = \frac{1}{1 + 4\gamma \eta^2 / \varrho g} \qquad 6.30)$$

Note that η may be expressed in the form

$$\eta = 2\pi(\nu \sin \theta)/c \qquad (6.31)$$

It can be seen that the transfer function G_p is a function of the liquid properties γ, c, and ϱ as well as the frequency ν of the acoustic wave and the angle of incidence θ. The quantity

$$2(\nu \sin \theta)/c = 2 \sin \theta/\Lambda \equiv 1/d \qquad (6.32)$$

is the spatial frequency of the interference pattern at the liquid surface, hence the transfer function varies as the inverse square of the spatial frequency at frequencies such that

$$4\gamma\eta^2 \gg \varrho g \qquad (6.33)$$

Forms of transfer functions other than that given in Eq. (6.30) may be used to express the relationship between the ripple amplitude A and the radiation pressure Π, or the ripple amplitude A and the intensity I. Normalized, these relationships all reduce to

$$G = [1 + (4\gamma\eta^2/\varrho g)]^{-1} \qquad (6.34)$$

where the normalized transfer function G may be interpreted as the G_p of Eq. (6.30), as

$$G_\pi = \varrho g \frac{A}{\Pi} \qquad (6.35)$$

or as

$$G_I = \frac{2\varrho g}{c} \frac{A}{I} \qquad (6.36)$$

6.4. INTERACTION OF LIGHT WITH THE LIQUID SURFACE

The purpose of this section is to demonstrate that when coherent light is reflected from a liquid surface having a configuration described by Eq. (6.18) the light will be so affected as to form an image of the acoustical field in the object. To do this it will be sufficient to show that the amplitude of the light wave has the same spatial distribution, over some region in space, as the amplitude of the sound wave in the object.

The incident light wave may be characterized by the equation

$$V(y, z) = D \exp i[(\eta_l y + \zeta_l z)] \qquad (6.37)$$

which describes a plane wave of constant amplitude D incident at an angle θ_l to the liquid surface. The wavelength of the light wave is λ. Furthermore

$$\eta_l = (2\pi/\lambda) \sin \theta_l \qquad (6.38)$$

and

$$\zeta_l = (2\pi/\lambda) \cos \theta_l \qquad (6.39)$$

To avoid confusing the coordinate z with the $z(X_i, Y_i, y)$ of Eq. (6.18), a function h is defined such that

$$h(X_i, Y_i, y) = z(X_i, Y_i, y) \qquad (6.40)$$

For sufficiently small values of θ_l (i.e., for light incident at near normal to the surface) the difference in travel distance between reflection from the surface $z = 0$ and the surface $z = h$ is $2h$. The reflected wave should be characterized by the equation

$$V(X_i, Y_i, y, z) = R \exp(i\{\eta_l y + \zeta_l[z - 2h(X_i, Y_i, y)]\}) \qquad (6.41)$$

where R is the amplitude of the reflected light which is always less than the incident amplitude D. Substituting for $h(X_i, Y_i, y)$ from Eq. (6.18) and rearranging yields the expression

$$V(X_i, Y_i, y, z) = R \exp(2i\zeta_l B) \exp[4i\zeta_l A \cos(2\eta y + \phi)] \exp[i(\eta_l y + \zeta_l z)] \qquad (6.42)$$

There is an identity, namely,

$$\exp(i\sigma \cos \alpha) = \sum_{n=-\infty}^{\infty} i^n J_n(\sigma) \exp(-in\alpha) \qquad (6.43)$$

which is useful for facilitating the physical interpretation of Eq. (6.42). Let

$$\sigma = 4\zeta_l A(X_i, Y_i) \qquad (6.44)$$

and

$$\alpha = 2\eta y + \phi(X_i, Y_i) \qquad (6.45)$$

Then Eq. (6.42) becomes

$$V(X_i, Y_i, y, z) = R \exp[i(\zeta_l z + 2B)] \sum_{n=-\infty}^{\infty} i^n J_n[4\zeta_l A(X_i, Y_i)]$$

$$\times \exp\{i[(\eta_l - 2n\eta)y - n\phi(X_i, Y_i)]\} \qquad (6.46)$$

Each value of n corresponds to a diffracted order of light. Normally the argument $4\zeta_l A(X_i, Y_i)$ is sufficiently small so that Eq. (6.46) can be approximated by

$$V(X_i, Y_i, y, z) = R \exp\{i\zeta_l[z + 2B(X_i, Y_i)]\} \exp(i\eta_l y)$$
$$+ 2i\zeta_l A(X_i, Y_i) \exp\{i[(\eta_l - 2\eta)y - \phi(X_i, Y_i)]\}$$
$$+ 2i\zeta_l A(X_i, Y_i) \exp\{i[(\eta_l + 2\eta)y + \phi(X_i, Y_i)]\}$$
$$\equiv V_0(X_i, Y_i, y, z) + iV_{+1}(X_i, Y_i, y, z)$$
$$+ iV_{-1}(X_i, Y_i, y, z) \qquad (6.47)$$

Consider the V_{+1} term which is

$$V_{+1} = 2R\zeta_l A(X_i, Y_i)\exp(i\{(\eta_l - 2\eta)y - \phi(X_i, Y_i) + \zeta_l[z + 2B(X_i, Y_i)]\})$$
(6.48)

Considering that $A(X_i, Y_i)$ is linearly related to $P_o(X_i, Y_i)$ through Eq. (6.19), there is a strong similarity between Eq. (6.5), representing the amplitude distribution of the acoustical wave in the object, and Eq. (6.48), representing the amplitude distribution of the optical wave, especially if $\eta_l = \eta$. The phase information is modified by the factor $B(X_i, Y_i)$ but the intensity distribution given by $V_{+1}V_{+1}^*$ is linearly related to the intensity distribution $U_o U_o^*$ in the object. Substituting for $A(X_i, Y_i)$ from Eq. (6.21) we find that

$$V_{+1}V_{+1}^* = \left(\frac{R\zeta_l P_r}{2\varrho\gamma c^2\eta^2}\right)^2 [P_o(X_i, Y_i)]^2$$
(6.49)

which shows that the acoustical intensity distribution is imaged as an optical intensity distribution.

6.5. EFFECTS PRODUCED BY PULSING THE SOUND WAVES

The previous sections of this chapter present an analysis of the characteristics of an acoustical hologram based upon the use of continuous waves. Although several experimenters [1,2] have been successful in demonstrating liquid-surface holography using continuous waves, considerable gain in effective surface flatness, surface stability, and spatial frequency response can be gained using bursts of sound approximately 100 wavelengths long. An analysis of the motion of a liquid surface when impinged upon by pulsed acoustical waves, which originate from within the liquid, will be given in this section.*

Take the particle motion of the acoustical wave to be in the y–z plane and for simplicity omit any consideration of variations in the x-direction. The pressure balance equation is similar to Eq. (6.15) for the continuous wave case except that terms describing the effect of inertial forces, frictional forces, and the intermittent nature of the radiation pressure must be added. The intermittent nature of the radiation pressure can be handled by use of

---

* The analysis relating to intermittent or pulsed acoustical radiation was first derived by T. J. Bander in proprietary internal technical reports, Battelle-Northwest, Richland, Washington.

a unit step function which has the property that

$$u(t) = 0 \quad \text{for} \quad t < 0$$
$$= 1 \quad \text{for} \quad t > 0$$

Furthermore, if the identity

$$\cos 2x = 2 \cos^2 x - 1$$

is used and if the acoustic pressure amplitudes P_o and P_r are assumed equal, Eq. (6.17) becomes

$$\Pi = \frac{4I_a}{c} \, [u(t) - u(t - \varDelta)] \cos^2 \eta y$$

where $\varDelta$ is the width of the acoustic pulse. The phase factor ϕ has been eliminated since only a single domain in the hologram is being considered and ϕ is thus a constant which can, for purposes of this analysis, be set equal to zero.

Frictional forces arise because of the fluid's viscosity. The shearing stress on the left face of the volume element shown in Fig. 6.4 can be written

$$\tau_L = \eta_v \frac{\partial}{\partial y} \left(\frac{\partial z}{\partial t} \right) \tag{6.50}$$

where η_v is the shear-viscosity coefficient. The shearing stresses on the right face of the element, τ_R are

$$\tau_R = \tau_L + \frac{\partial \tau_L}{\partial y} \, dy$$
$$= \eta_v \left[\frac{\partial}{\partial y} \left(\frac{\partial z}{\partial t} \right) + \frac{\partial^2}{\partial y^2} \left(\frac{\partial z}{\partial t} \right) dy \right] \tag{6.51}$$

The vertical force developed as a result of these two shearing stresses is

$$\tau_R \left(z + \frac{\partial z}{\partial y} \, dy \right) - \tau_L z$$
$$= \eta_v \left[z \frac{\partial^2}{\partial y^2} \left(\frac{\partial z}{\partial t} \right) dy + \left(\frac{\partial z}{\partial y} \right) \left(\frac{\partial^2 z}{\partial y \, \partial t} \right) dy + \left(\frac{\partial z}{\partial y} \right) \left(\frac{\partial^3 z}{\partial y^2 \, \partial t} \right) (dy)^2 \right] \tag{6.52}$$

Since the final term in the brackets is second order in dy, it can be neglected in comparison to the first two terms. Thus the force contributed by the

Fig. 6.4. The volume element of unit depth in the x-direction has the radiation pressure Π acting upon the surface, a gravitational pressure $\varrho g z$ acting upon its mass, a pressure $\gamma\, \partial^2 z/\partial y^2$ due to surface tension acting upon the surface, a shear force τ_L acting on the left-hand side of height z, and a shear force τ_R acting on the right-hand side. The sum of these forces must equal the inertial force.

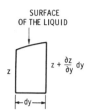

SURFACE OF THE LIQUID

$z + \frac{\partial z}{\partial y}\, dy$

z

$\leftarrow dy \rightarrow$

shearing stresses is

$$\eta_v\left[z\, \frac{\partial^3 z}{\partial y^2 \partial t}\, dy + \frac{\partial^2 z}{\partial y \partial t}\, \frac{\partial z}{\partial t}\, dy\right] = \eta_v\, \frac{\partial}{\partial y}\left[z\, \frac{\partial^2 z}{\partial y \partial t}\right] dy \qquad (6.53)$$

The sum of all the forces acting upon a volume element must be equal to the time rate of change of momentum. Thus the governing equation becomes

$$\frac{4I_a}{c}\,[u(t) - u(t - \Delta)]\cos^2 \eta\gamma + \gamma\, \frac{\partial^2 z}{\partial y^2}$$

$$+ \eta_v\, \frac{\partial}{\partial y}\left[z\, \frac{\partial^2 z}{\partial y \partial t}\right] - \varrho g z = \varrho\, \frac{\partial}{\partial t}\left(z\, \frac{\partial z}{\partial t}\right) \qquad (6.54)$$

This is a nonlinear differential equation which is difficult to analyze except by numerical methods. In order to deal with this equation using Fourier and Laplace transform methods, it must be linearized. If the assumption is made that the inertial and viscous forces are acting on a volume element of constant height h rather than on a volume of height z, Eq. (6.54) will be linearized in the form

$$\frac{4I_a}{\varrho c}\,[u(t) - u(t - \Delta)]\cos^2 \eta\gamma + \frac{\gamma \partial^2 z}{\varrho \partial y^2} + \frac{\eta_v h}{\varrho}\, \frac{\partial^3 z}{\partial y^2 \partial t} - g z - h\, \frac{\partial^2 z}{\partial t^2} = 0$$

$$(6.55)$$

At the time $(t = 0)$ of the initiation of the acoustical pulse, the surface is at rest $(\partial z/\partial t = 0)$ at its quiescent level $(z = 0)$. The Laplace transform of Eq. (6.55) is

$$\frac{4I_a}{\varrho c}\left[\frac{1 - \exp(-\Delta s)}{s}\right] + \frac{\gamma}{\varrho}\, \frac{\partial^2 \bar{z}}{\partial y^2} + \frac{\eta_v h}{\varrho}\, s\, \frac{\partial^2 \bar{z}}{\partial y^2} - g \bar{z} - h s^2 \bar{z} = 0 \quad (6.56)$$

where

$$\bar{z} = \int_0^\infty \exp(-st)\, z\, dt \qquad (6.57)$$

Rearranging terms, we can write

$$\frac{\partial^2 \bar{z}}{\partial y^2} - \frac{\varrho(g + hs^2)}{(\gamma + \eta_v hs)} \bar{z} + \frac{4I_a}{c} \left[\frac{1 - \exp(-\varDelta s)}{s(\gamma + \eta_v hs)} \right]$$

$$+ \frac{4I_a}{c} \left[\frac{1 - \exp(-\varDelta s)}{s(\gamma + \eta_v hs)} \right] \cos^2 \eta_y = 0 \qquad (6.58)$$

The solution of this differential equation consists of a complimentary solution plus a particular solution, i.e.,

$$\bar{z} = \bar{z}_c + \bar{z}_p \qquad (6.59)$$

The complimentary solution is given by the expression

$$\bar{z}_c = A_1 \sinh \left[\sqrt{\frac{p(g + hs^2)}{\gamma + \eta_v hs}} \, y \right] + A_2 \cosh \left[\sqrt{\frac{\varrho(g + hs^2)}{\gamma + \eta_v hs}} \, y \right] \qquad (6.60)$$

where A_1 and A_2 are constants whose values are determined by the boundary conditions. Finding the inverse transform of the complimentary solution will not be pursued. Instead, we focus attention upon the particular solution which we will assume to be of the form

$$\bar{z}_p = A_3 \cos^2 \eta y + A_4 \qquad (6.61)$$

where A_3 and A_4 are constants with respect to y. The values of A_3 and A_4 are determined in the following manner. Using the assumed form for $\bar{z}_p$, Eq. (6.58) becomes

$$- \left[4\eta^2 A_3 - \frac{\varrho(g + hs^2)}{(\gamma + \eta_v hs)} A_3 + \frac{8I_a}{cs} \frac{1 - \exp(-\varDelta s)}{(\gamma + \eta_v hs)} \right] \cos^2 \eta y$$

$$+ \left[2\eta^2 A_3 - \frac{\varrho(g + hs^2)}{\gamma + \eta_v hs} A_4 \right] = 0 \qquad (6.62)$$

Since this equation must hold for any value of y, the bracketed terms must be independently equal to zero, hence the values of A_3 and A_4 can be readily determined to be

$$A_3 = \frac{8I_a[1 - \exp(-\varDelta s)]}{\varrho chs[(s + d_1)^2 + d_2^2]} \qquad (6.63)$$

and

$$A_4 = \frac{16\eta_v \eta^2 I_a(d_4 + s)[1 - \exp(-\varDelta s)]}{\varrho^2 chs(d_3^2 + s^2)[(s + d_1)^2 + d_2^2]}$$

where

$$d_1 = \frac{2\eta^2}{\varrho} \eta_v \tag{6.64}$$

$$d_2{}^2 = \frac{4\eta^2(\varrho\gamma - h\eta_v{}^2\eta^2) + \varrho^2 g}{\varrho^2 h} \tag{6.65}$$

$$d_3{}^2 = \frac{g}{h} \tag{6.66}$$

and

$$d_4 = \frac{\gamma}{\eta_v h} \tag{6.67}$$

Eq. (6.61) can now be written in the form

$$\bar{z}_p = \left(\frac{8I_a}{\varrho ch}\right) \frac{[1 - \exp(-\Delta s)]\cos^2 \eta y}{s[(s + d_1)^2 + d_2{}^2]}$$

$$+ \left(\frac{16\eta_v\eta^2 I_a}{\varrho^2 ch}\right) \frac{(d_4 + s)[1 - \exp(-\Delta s)]}{s(d_3{}^2 + s^2)[(s + d_1)^2 + d^2{}_2]} \tag{6.68}$$

The only remaining step toward the goal of finding the displacement of the liquid surface as a function of time and position is to take the inverse Laplace transform of $\bar{z}_p$. When this is done the result, which applies only for $t > \Delta$ is,

$$z_p = \frac{4I_a \exp(-d_1 t)}{c(4\gamma\eta^2 + \varrho g)} [2\cos^2 \eta y - 1]$$

$$\times \left\{\exp(d_1 \Delta)\left[\frac{d_1}{d_2} \sin d_2(t - \Delta) + \cos d_2(t - \Delta)\right] - \frac{d_1}{d_2} \sin d_2 t + \cos d_2 t\right\}$$

$$+ \frac{4I_a}{\varrho cg} \{\cos d_3(t - \Delta) - \cos d_3 t\} \tag{6.69}$$

It is instructive to split z_p of Eq. (6.69) into two parts; part, z_{pb}, describes the displacement associated with the overall bulge on the liquid surface and the other, z_{pr}, describes the displacement associated with the ripple. Thus the total displacement, z_p, is given by

$$z_p = z_{pb} + z_{pr} \tag{6.70}$$

i.e.,

$$z_{pr} = \frac{4I_a \exp(-d_1 t)}{c(4\gamma\eta^2 + \varrho g)} [2\cos^2 \eta y - 1]$$

$$\times \left\{\exp(d_1 \Delta)\left[\frac{d_1}{d_2} \sin d_2(t - \Delta) + \cos d_2(t - \Delta)\right] - \frac{d_1}{d_2} \sin d_2 t - \cos d_2 t\right\}$$

$$\tag{6.71}$$

and

$$z_{pb} = \frac{4I_a}{\varrho cg} \{\cos d_3(t - \varDelta) - \cos d_3 t\} \tag{6.72}$$

Using the parameter values of Table 6.1 yields the following:

$$B = \frac{4I_a}{\varrho cg} = 2.7 \times 10^{-2} \text{ cm} \tag{6.73}$$

$$2A = \frac{4I_a}{c(4\gamma\eta^2 + \varrho g)} = 7.0 \times 10^{-5} \text{ cm} \tag{6.74}$$

The quantities A and B are defined just as they were in Eqs. (6.26) and (6.27) for the continuous wave case. The ratio $B/2A$ is large but in the pulsed case,

Table 6.1. Numerical Results for Water at 25° C

Properties of the Liquid

    ϱ, density—1 g cm^{-3}
    γ, surface tension—72 dyne cm^{-1}
    η_v, viscosity—0.009 dyne sec cm^{-2}
    c, velocity of sound—1.5 × 10^5 cm sec^{-1}

Operational Parameters

    θ, angle of incidence—5 deg
    ν, frequency of the sound—10^7 Hz
    I_a, intensity of the sound—10^6 erg sec^{-1} cm^{-2}

Constants

    g, acceleration of gravity—980 cm sec^{-2}
    h, effective vertical displacement of the water surface
        (taken to be $2I_a/cg = 1.36 \times 10^{-2}$ cm)

Results

$d_1 = 24$ sec^{-1}	$t_1 = 1/d_1 = 4.2 \times 10^{-2}$ sec
$d_2 = 5.3 \times 10^3$ sec^{-1}	$t_2 = 1/d_2 = 1.9 \times 10^{-4}$ sec
$d_3 = 2.7 \times 10^2$ sec^{-1}	$t_3 = 1/d_3 = 3.7 \times 10^{-3}$ sec
$d_4 = 5.9 \times 10^5$ sec^{-1}	$t_4 = 1/d_4 = 1.7 \times 10^{-6}$ sec

$t_1 = 1/d_1$, time to damp ripple oscillation to $1/e$ of initial value; $t_1 = 0.042$ sec

$f_r \equiv \dfrac{d_2}{2\pi}$, frequency of the ripple oscillation; $f_r = 845$ Hz

$f_b \equiv \dfrac{d_3}{2\pi}$, frequency of oscillation of the bulge; $f_b = 43$ Hz

z_{pb} and z_{pr} contain time-dependent factors in addition to the amplitudes A and B. These time-dependent factors may be used to modify z_{pb} and z_{pr} so that the ratio z_{pb}/z_{pr} is much more favorable than the ratio $B/2A$. If, for instance, the hologram is illuminated for a relatively short time interval centered on the time at which maximum ripple amplitude, z_{pr}, occurs, the ratio z_{pb}/z_{pr} has a value of only 1.56 for a pulse width $K = 80 \ \mu\text{sec}$. To show this, set

$$\frac{\partial z_{pr}}{\partial t} = 0 \tag{6.75}$$

from which it may be deduced that

$$\tan d_2 t = \frac{\sin(d_2 \varDelta)}{\cos(d_2 \varDelta) - \exp(-d_1 \varDelta)} \tag{6.76}$$

Using 80×10^{-6} sec for the pulse width $\varDelta$ and the values from Table 6.1 for d_1 and d_2,

$$\tan d_2 t = -4.74 \tag{6.77}$$

or

$$t = 3.35 \times 10^{-4} \text{ sec} \tag{6.78}$$

for the time at which the maximum ripple amplitude occurs. The values of z_{pb} and z_{pr} at this time are

$$z_{pr} = 1.46 \times 10^{-5} [2 \cos^2 \eta y - 1] \text{ cm} \tag{6.79}$$

and

$$z_{pb} = 2.3 \times 10^{-5} \text{ cm} \tag{6.80}$$

so that

$$z_{pb}/z_{prm} = 1.56 \tag{6.81}$$

where z_{prm} is the maximum value of z_{pr}. It is interesting to note that the maximum ripple amplitude occurs well after the pulse of sound has been terminated.

6.6 TRANSFER FUNCTION RELATING LIQUID-SURFACE RIPPLE DISPLACEMENT TO ACOUSTIC RADIATION PRESSURE

The properties of Fourier transforms may be applied to finding the spatial frequency response of a liquid surface. Suppose, for example, that the radiation pressure $\Pi(y)$ is not a function of just one single spatial frequency η as in Eqs. (6.17) and (6.71), but contains many spatial frequencies of different amplitudes such that it can be described by

$$\Pi(y) = \int_{-\infty}^{\infty} \hat{\Pi}(2\eta) \exp(-4\pi i \eta y) \, d\eta \qquad (6.82)$$

where $\hat{\Pi}(2\eta)$ is the amplitude of that component of the radiation pressure which has a spatial frequency 2η. The value of $\hat{\Pi}(2\eta)$ may be determined from

$$\hat{\Pi}(2\eta) = \int_{-\infty}^{\infty} \Pi(y) \exp(4\pi i \eta y) \, dy \qquad (6.83)$$

i.e., $\hat{\Pi}(2\eta)$ is the Fourier transform of $\Pi(y)$.

Corresponding to $\hat{\Pi}(2\eta)$ there will be a $\hat{z}(2\eta)$ which is the amplitude of the liquid-surface ripple displacement for the spatial frequency 2η. The transfer function for liquid-surface ripple displacement to acoustical radiation pressure is given by the ratio $\hat{z}/\hat{\Pi}$, the calculation of which may be begun by taking the Fourier transform of Eq. (6.54) wherein the term $(4I_a/c) \cos^2 \eta y$ will be replaced by $\Pi(y)$ of Eq. (6.82). This procedure yields

$$\hat{\Pi}[u(t) - u(t - \varDelta)] - 4\gamma\eta^2\hat{z} - 4\eta_v h\eta^2 \frac{\partial \hat{z}}{\partial t} - \varrho g\hat{z} - \varrho h \frac{\partial^2 \hat{z}}{\partial t^2} = 0 \qquad (6.84)$$

where

$$\hat{z} = \int_{-\infty}^{\infty} z \exp(-4\pi i \eta y) \, dy \qquad (6.85)$$

The next step is to solve Eq. (6.84) for $\hat{z}$ which can be done by applying the Laplace transform using the initial conditions $z = 0$ and $\partial z/\partial t = 0$ at $t = 0$. Eq. (6.84) becomes

$$\hat{\Pi}\left[\frac{1 - \exp(-\varDelta s)}{s}\right] - 4\gamma\eta^2\hat{\hat{z}} - 4\eta_v h\eta^2 s\hat{\hat{z}} - \varrho g\hat{\hat{z}} - \varrho h s^2\hat{\hat{z}} = 0 \qquad (6.86)$$

where

$$\hat{\hat{z}} = \int_{0}^{\infty} \hat{z} \exp(-st) \, dt \qquad (6.87)$$

Eq. (6.86) may be rearranged to the form

$$\hat{\hat{z}} = \frac{\hat{\Pi}[1 - \exp(-\varDelta s)]}{\varrho h s[(s + d_1)^2 + d_2{}^2]} \qquad (6.88)$$

where d_1 and d_2 are given by Eqs. (6.64) and (6.65). Thus, taking the inverse

Laplace transform of Eq. (6.88) we find that

$$\hat{z} = L^{-1}(\hat{z}) = \frac{\hat{\Pi} \exp(-d_1 t)}{\varrho h (d_1{}^2 + d_2{}^2)}$$

$$\times \left\{ \exp(d_1 t) \left[\frac{d_1}{d_2} \sin d_2 (t - \varDelta) + \cos d_2 (t - \varDelta) \right] - \frac{d_1}{d_2} \sin d_2 t - \cos d_2 t \right\}$$

$$(6.89)$$

The transfer function for the liquid-surface ripple amplitude to the radiation pressure is

$$G = \frac{\hat{z}}{\hat{\Pi}} = \frac{\exp(-d_1 t)}{\varrho g + 4 \gamma \eta^2}$$

$$\times \left\{ \exp(d_1 \varDelta) \left[\frac{d_1}{d_2} \sin d_2 (t - \varDelta) + \cos d_2 (t - \varDelta) \right] - \frac{d_1}{d_2} \sin d_2 t - \cos d_2 t \right\}$$

$$(6.90)$$

Equation (6.90) contains a wealth of detailed information about the response of a liquid surface to a pulse of acoustical energy having a spatial

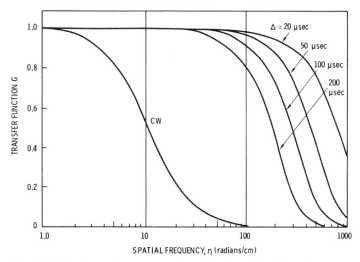

Fig. 6.5. The dynamic response characteristics of liquid-surface holograms, as calculated from Eq. (6.90), for the special case $t = \varDelta$ are plotted in this figure. It is apparent from this plot that a liquid surface acts as a low-pass filter and that the cut-off frequency increases as the pulse width $\varDelta$ decreases. The spatial frequency at cut-off (for a pulse width $\varDelta = 20\ \mu\text{sec}$) is almost two orders of magnitude greater than the cut-off spatial frequency of a continuous wave.

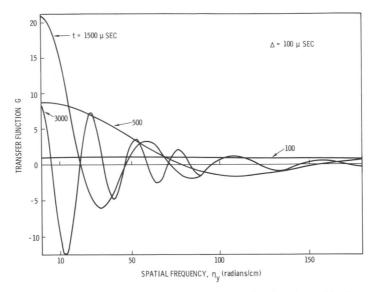

Fig. 6.6. The behavior of the liquid-surface transfer function with time of interrogation as a parameter is shown in this figure. Note that it becomes oscillatory indicating that the ripple amplitude reverses in sign periodically.

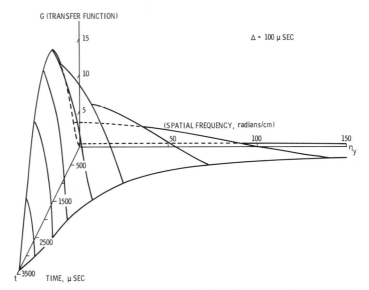

Fig. 6.7. This figure describes the transfer function as a surface with time of interrogation and spatial frequency as variables. Note that gain is increased relative to $t = \Delta$ for $t > \Delta$ and reaches a maximum at $t \cong 1500\ \mu\text{sec}$. A continuation of the figure beyond the curved line on the $t-\eta$ axis would yield negative values of G. Even further extension reveals an oscillatory surface.

frequency 2η, and a pulse width Δ. A full portrayal of the transfer function G would require a four-dimensional plot. The four axes would be G the transfer function; 2η the spatial frequency; Δ the pulse width and the time at which the transfer function is to be evaluated.

Since we cannot draw four-dimensional diagrams we must resort to several lower-dimensional plots. The first and simplest of these is shown in Fig. 6.5. This figure dramatically portrays the increase in performance as the acoustic pulse length is decreased from the continuous wave case. The light pulse occurs at the end of the acoustic pulse. Note that the spatial bandwidth (for water) increases from about 8 rad/cm for continuous wave

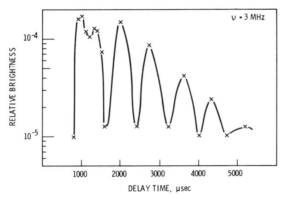

Fig. 6.8. The experimental data plotted in this figure show the relative brightness of first-order diffracted light as a function of the delay between the time of initiation of the acoustical pulse and the occurrence of the laser pulse used to illuminate the hologram. The experimental arrangement was such that the travel time of the acoustical wave from the transducer to the hologram was 810 μsec. The pulse width Δ of the acoustical wave was 100 μsec and that of the laser was 50 μsec. By increasing the laser delay time, the liquid surface is sampled for 50 μsec intervals at times increasingly delayed with respect to the time that the acoustic pulse is applied. Note that the maximum brightness occurs 100 μsec after the acoustical pulse has been terminated. Acoustical pulses occurred at a rate of 100 per second. Note also that the ripple structure in the hologram is oscillating as is indicated by periodic variations in the brightness of the diffracted light. These variations have an average period of approximately 825 μsec. Furthermore the oscillations are decaying to 37% of their maximum intensity in about 2 msec.

to 600 rad/cm for a 20 μsec pulse. This plot alone explains why the initial experiments using continuous wave yielded rather poor results.

Figure 6.6 is a plot of the response, with time at which the light pulse interrogates the surface as a parameter. The acoustic pulse length is maintained at a constant 100 μsec. Note that the transfer function becomes oscillatory.

Figure 6.7 is a three-dimensional plot of transfer function with spatial frequency and time of interrogation as variables. The transfer function is normalized to unity for $t = \Delta$. We have plotted results in the first positive region. If it were continued, we would find that the transfer function goes negative as shown in Fig. 6.6. Note that a maximum exists, for $t > \Delta$ but the bandwidth to the first zero is decreased.

In electronic engineering, a quantity often used as a quality criterion is the gain–bandwidth product. If we perform a simple calculation of this type (calculate the area under a cross section of the curve parallel to the $G-\eta_y$ plane) we find that a maximum occurs at $t = 400$ μsec for this particular example. That is, an optimum image in some sense, is obtained when interrogation of the surface occurs 300 μsec after the acoustical pulse has been turned off.

An experimental verification of the variation in image brightness as a function of time of interrogation is shown in Fig. 6.8. The maxima correspond to the maxima and minima of G and the minima to the zeros of G along the t-axis in Fig. 6.7.

REFERENCES

1. R. K. Mueller and N. K. Sheridan, *Appl. Phys. Lett.* **9**:328–329 (1966).
2. P. S. Green, *Acoustic Holography*, LMSC Tech. Rept. 6-77-67-42 Sept., 1967, Lockheed Missiles and Space Company, Palo Alto, California.

Chapter 7

Other Acoustical Holography Methods

As we have mentioned earlier, the main problem in the development of practical devices utilizing holographic principles is the spatial detection and recording system. We have described in great detail the two methods that are furthest along in the development process. In this chapter we discuss a number of methods that have appeared in the literature and some which we have worked on.

7.1. PHOTOGRAPHIC FILM

The use of photographic film for the recording of acoustical holograms appears to be the first published reference to the application of holography to acoustical imaging [1]. This method relies on the fact that ultrasonic stimulation of photographic emulsion before or during development sensitizes the film. Therefore, if the acoustical field has a spatial intensity variation, those parts of the emulsion suffering the greatest agitation will develop faster than other regions. Hence, when the plate is developed, a hologram results.

In general, this method requires high acoustic intensities. P. Greguss, the main exponent of this method, devised certain procedures to sensitize the emulsion. First, he found that pre-exposure to light was essential and second, he immersed the plate in a dilute solution of thiosulfate. He achieved acoustical exposure times of 20–50 sec at intensities (in each beam) of 0.5 W/cm². Although he obtained interference patterns, they were never of a quality sufficient to reconstruct images. This method, then, although it works in principle, has not proved to be successful.

Other variations in the use of photographic emulsions for recording ultrasound involve acoustical stimulation of the film during development [2], and fixing [3]. A related technique uses a starchcoated plate in iodine solution [4]. The relative efficiencies of these methods are summarized in Table 7.3 at the end of this chapter.

7.2. ULTRASOUND CAMERA

The ultrasound camera is the acoustical equivalent of the image orthicon used in television systems. It seems that this idea was first proposed by Sokolov in 1937 [5], but was not fully developed until quite recently [6]. The photosensitive face of the image orthicon is replaced by a sound-sensitive surface, usually of resonant quartz but sometimes of barium titanate. The sound field imparts an oscillating stress to the piezoelectric face-plate which, in turn, generates an alternating voltage across the plate. An electron beam, scanning across the back of the face-plate causes secondary emission in proportion to the voltage amplitude, modulated at the frequency of the sound wave. The secondary electrons are collected and the current is used to modulate the intensity of an oscilloscope trace which is synchronized with the electron beam in the camera tube.

Since we want the display to produce a visual image of the acoustic intensity variations across the camera face, it is crucial that there be an

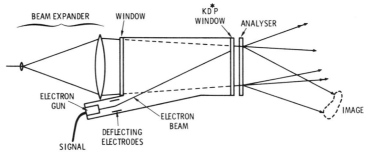

Fig. 7.1. The linearly polarized light of the laser is elliptically polarized by the KD*P crystal in direct proportion to the electron beam intensity. The analyzer then allows only the linearly polarized light to pass. Since the electron beam intensity represents the acoustical hologram, the remaining light forms an image.

accurate spatial translation from acoustic pressure to potential distribution. An exact translation requires that each point on the face-plate acts independently of all neighboring points. This is, of course, not possible since the face-plate must have a finite thickness, both for reasons of strength (it must support a vacuum) and for generation of an adequate potential. The face-plate is generally designed as a resonant element, $\lambda/2$ in thickness. The resulting intrinsic resolution becomes, approximately, $\lambda/2$ [7]. The wavelength referred to here is that in the material of which the face-plate is constructed. Therefore, in terms of the resolution at the object in water the intrinsic resolution capability of the face-plate becomes approximately $2\lambda_w$ where λ_w is the wavelength in water [8].

An additional constraint on the resolution of the ultrasound camera is the limited field of view, arising because of the low critical angle at the water–quartz interface. This effectively limits the angular aperture of the tube to about $10°$ with the attendant lowered resolution. Jacobs has estimated that the attainable resolution, including this effect, is approximately $3\lambda_w$ [8].

The second part that is required to make the acoustic camera a complete imaging system is a display. For traditional use, the electronic signal derived from the camera is used to drive a monitor screen. This can also be used for holography, together with a camera for photographing the screen. The negative then becomes the hologram from which an image may be reconstructed. To take full advantage of the extremely fast electronic scan mechanism of the camera, however, we should use a real time display such as suggested earlier.

A particular example of such a display is described by Boutin, Marom, and Mueller [9]. The display consists of a cathode-ray tube with a KD*P crystal on the face as shown in Fig. 7.1. The KD*P crystal acts as a polarizer with polarization dependent upon the electronic charge deposited on its surface. Since the electron beam deposits charge which is spatially varying according to the acoustic intensity on the face of the ultrasound camera, we have a representation of the acoustic field in terms of polarization. In its inactive state, the KD*P crystal is linearly polarized. The analyzer is oriented with its axis of polarization at right angles to the KD*P axis of polarization. If laser light, also linearly polarized parallel to that of the KD*P crystal, is introduced along the optical axis, no light will emerge from the analyzer. When the crystal is activated by the electrical field it causes elliptical polarization of the light being transmitted. A portion of the light will then pass the analyzer with the result that the optical equivalent of the acoustic field appears at the output side of the analyzer. This optical field, of course, will form an image if the acoustic field is a hologram.

The ultrasound camera and KD*P display system seem to answer all the problems for development of a real-time holography system. Indeed, for small fields of view it does. The major drawback is that of fabrication of the sensitive crystal for the camera. As we have seen, the resolution of the camera is inversely proportional to the thickness of the crystal. Unfortunately, this also means that the mechanical strength of the crystal becomes less with decreased thickness, which, because it must support a vacuum, requires a decreased aperture. A number of designs for improving this aspect of the camera have been proposed and are described by Jones and Miles [7]. Even if these methods succeed and larger apertures are built, the effective aperture is still limited by the critical angle problem. Therefore, it seems likely that this system will be useful mainly for low-resolution laboratory work.

7.3. PARTICLE CELL

We might have headed this section "Pohlman Cell," since our work is based on his original concept [10]. However, in our work we extended his ideas, and used sufficiently different techniques and materials to justify the above title.

The Pohlman Cell was first suggested in 1939 [10] and described in detail in a Dutch patent in 1940 [11]. This form of ultrasound detector, shown in Fig. 7.2, consists of a sandwich containing a metallic powder in suspension in a suitable liquid. In Pohlman's configuration, one side of the sandwich is a glass plate while the other, exposed to the sound field is a thin membrane. The suspension consisted of aluminum flakes in xylol, although in the patent Pohlman suggests also V_2O_5 (vanadium pentoxide) and Fe_2O_3

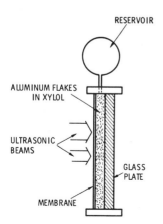

Fig. 7.2. The original particle cell as outlined by Pohlman. Acoustical energy incident from the left aligns the particles perpendicularly to the field, thus presenting a reflective aspect to light.

(ferric oxide). When the suspension is ultrasonically stimulated, the Al flakes orient themselves so as to present maximum area to the sound field. If the suspension is illuminated by light, one sees, on reflection, the acoustic intensity pattern impinging on the cell, since areas of maximum intensity will present a more reflective surface to the light.

In the case of V_2O_5 or Fe_2O_3, which are more or less transparent to light, the alignment of the particles tends to make the cell a polarizer. The use of suitably polarized light source makes it possible to visualize the sound field.

In another version of the cell, Pohlman suggests the use of colored particles suspended in a cell placed horizontally. The particles will tend to fall out of suspension to form a uniform color. When stimulated by a sound field the particles will be disturbed and tend to collect in the regions of low intensity. When the sound is turned off, we are left with a picture of the sound field.

All of the variations on the Pohlman cell offer approximately the same advantages and disadvantages. One advantage is the high intrinsic resolution which comes about because of the very small size of the particles in the suspension. Since they are much smaller than the wavelength of sound (at least for 10 MHz), and the coupling membrane is also much less than a wavelength, there is no intrinsic limitation on resolution.

The real limitation of this detection method is the exposure and relaxation time for the cell. Although the sensitivity is high, in the sense that very small energies are required to orient the particles, the response time is too slow for real-time display of holograms (seconds).

Our work with the particle cell has been mostly experimental. We have attempted to optimize the particle solvent mixture to obtain fast response time with good image formation. One of the first things we found was that the major action was not particle orientation but particle migration. That is, when two plane acoustic waves interfere in the cell, there is a migration of particles to areas of low intensity, forming a linear spatial grating. Furthermore, there appeared to be a constant circulation of particles along the lines of the grating. This suggested to us that the cell should, therefore, be used in transmission rather than reflection. Consequently, we built the test system shown in Fig. 7.3. This system has the drawback that the light must propagate through the water. However, in the laboratory one can maintain clear water, and the versatility of the system for testing different cells overcomes the objection.

As we mentioned earlier, our approach has been primarily experimental. This is so because the problem is an extremely complicated one,

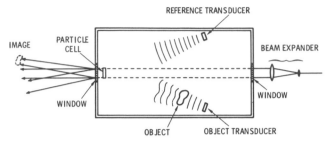

Fig. 7.3. Acoustical tank for the testing of various particles and solvents for an acoustical particle cell detector.

involving fluid dynamics, electric field theory, and surface chemistry. Since we did not have the luxury of time or money, we decided to follow certain clues in the literature on an experimental basis.

A particularly interesting series of papers came to light in a literature survey, indicating that the dipole moment of the suspension is an important factor in recovery time [12–19]. The original purpose of the work was the study of the diffusion of particles in colloidal solutions. The studies involved the use of ultrasonic standing waves to diffract the light into various orders which were then measured. It was observed that for certain solutions, the diffracted orders persisted after the ultrasound was turned off. This observation triggered the series of investigations of this phenomenon, described in the mentioned references.

Table 7.1. Experimental Reaction Times for a Particle Cell Consisting of Aluminum 560 Powder in Several Solvents[a]

Solvent	Dipole moment/viscosity ratio	Reaction time, sec
Acetonitrile	9.8	2
Acetone	9.1	4
Methyl alcohol	2.9	6
Water (soapy)	1.8	6
Ethyl alcohol	1.4	8

[a] Reaction time is defined as the time it takes for a hologram image to reform after the object is moved. The inverse relationship between the dipole moment-viscosity ratio to reaction time is clearly evident.

Table 7.2[a]

Chemical	Formula	Density ϱ, g/cm³	Viscosity η, centipoises	Dipole moment μ, debyes	Dipole moment/viscosity ratio
Acetic acid	$C_2H_4O_2$	1.049	1.30	1.92	1.48
Acetone	C_3H_6O	0.791	0.316	2.86	9.05
Acetonitrile	C_2H_3N	0.786	0.345	3.39	9.85
Benzene	C_6H_6	0.879	0.652	0.10	0.153
Carbon tetrachloride	CCl_4	1.595	0.959	0	0
Ethyl alcohol	C_2H_6O	0.789	1.20	1.73	1.44
Ethylene dichloride	$C_2H_4Cl_2$	1.256	0.800	2.94	3.68
Ethylene glycol	$C_2H_6O_2$	1.109	19.9	3.59	0.18
Freon	$C_9HF_{17}O_3$	1.723	2.2	—	—
Heptane	C_7H_{16}	0.684	0.409	~0	0
Mesitylene	C_9H_{12}	0.864	—	0	0
Methyl alcohol	CH_4O	0.791	0.597	1.71	2.86
N,N-dimethylformamide	C_3H_7NO	0.945	—	3.85	—
p-Xylene	C_8H_{10}	0.87	0.648	0.1	0.154
Water	H_2O	1.000	1.002	1.82	1.82

[a] Various solvents that were tested. These are shown to illustrate the great variety of solvents available. In general, those with high moment-viscosity ratio were superior.

It was soon decided that the reason for the persistance of the diffraction pattern was that the particles in solution had migrated to the nodal planes of the standing wave. It was shown, however, that the concentration of particles was too low to make a significant change in index of refraction, so that diffraction should not occur on that account. It was also noted that for some solvents, even though the particles were aligned to form a grating, diffraction did not occur, thus proving that something else was at work. The theory subsequently proposed and supported by experiments was that the particles, electrically charged, attract the free dipoles present in polar liquids. This division of the liquid into regions of dipole concentrations results in a medium of varying refringence index large enough to diffract light [16].

The decay of the presence of diffraction after the ultrasound is turned off was experimentally determined to follow the exponential law

$$I_p = I_{0p} \exp(-kt)$$

where k is dependent upon the solvent, the material in suspension and its concentration. The final paper in the series links the relaxation time of the light diffracted by the grating to the laws of diffusion with good agreement.

From this series of papers one can form a tentative theory and set the direction of experimentation for making a practical, particle-cell holography system. First of all, we know that a solvent with high-dipole moment is required. Then, one needs to experiment with different particulates to find a combination of solvent and particle concentration for maximum diffraction efficiency. At the same time, we require fast diffusion of the particles when the sound field changes so that the diffraction pattern will change rapidly. This requires a solvent of low viscosity.

Our limited experimental work gave the results listed in Table 7.1. These results pertain to the use of aluminum 560 powder in a cell with 0.001-in.-thick mylar sides spaced by 5 mm. The solvents listed in Table 7.2 were also tried but were inferior to those listed in Table 7.1. Figure 7.4 shows some reconstructions using different particles in soapy water and Fig. 7.5 represents our best reconstruction to date.

Although we have not succeeded in designing a particle cell with sufficient reaction time for real-time viewing of objects in motion, we feel that the knowledge to do so is available. A concentrated effort applied to developing the theory along with a sound experimental program should yield results.

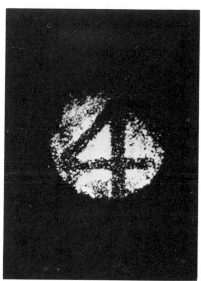
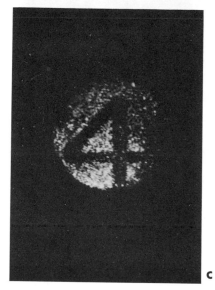

Fig. 7.4. The three photographs represent images of acoustical holograms formed in a particle cell. The solvent was soapy water, and (a) the particles were Aluminum flakelets 1–10 μm in diameter and thickness less than 100 nm in a solution ratio of 5 ml/300 ml, (b) Pearl Afflair flakelets 3–50 μm in diameter and thickness less than 100 nm in a solution ratio of 1 ml/100 ml, and (c) tipersil needles 1 μm in diameter and length of 10–20 μm in a solution ratio of 2ml/100 ml, respectively. The object was a brass figure 1.25 mm thick, 4.5 cm tall and 2.86 cm wide seen in transmission. The acoustical frequency was 5 MHz.

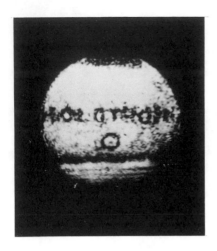

Fig. 7.5. Our best particle cell hologram yielded this image of the word HOLO-TRON engraved in a 6.3-mm-thick aluminum plate. The cell medium was aluminum flakelets in soapy water, the frequency 5 MHz, the letters were 0.25 mm deep with line widths of 0.5 mm.

7.4. BRAGG DIFFRACTION IMAGING

Although Bragg diffraction imaging is not holography, it is lensless, preserves phase and amplitude, and is accomplished in real time. Consequently, it appears to provide all the advantages of holography and for this reason we discuss it here.

7.4.1. Bragg Reflection

Consider the reflection of a plane light wave from a series of equally spaced reflecting surfaces as shown in Fig. 7.6a. If the reflected light is to be plane also, we must have the path length difference between the rays equal to a multiple of the light wavelength. By geometrical manipulations we find that this requires that the angle ψ be specified by the relation

$$\sin \psi = \pm \frac{n\lambda}{2\Lambda} \tag{7.1}$$

where n is an integer and Λ is the separation of the surfaces. This angle, known as the Bragg angle, results in a perfect plane wave emerging from the mirror grating. If the angle of the entering wave deviates from the Bragg angle, the emerging light will have phase deviations resulting in interference effects when the intensity is observed.

We next consider the case of a cylindrical optical wave reflected by a curved set of mirrors as shown in Fig. 7.6b. In this case we require that the reflected wave have a cylindrical wave front. The geometry is more complicated for this case, but the answer turns out to be the same. That is, the light ray must hit each cylindrical surface at the Bragg angle evaluated in Eq. (7.1). The points on the cylindrical surface which fulfill this requirement can be shown to lie on a cylinder passing through the origin of the light waves and the center of the cylindrical reflecting surfaces [20]. Thus, we see that the reflected light wave represents a virtual line source which also lies on the same cylinder as the real light source and the center of the cylindrical mirror system. Note that only one line of reflection on each cylinder contributes to the virtual source wave. Light reflecting from other parts of the cylinder contributes "noise."

The next step in the generalization of Bragg reflection involves spherical reflecting surfaces centered on some point and spherical light waves. In order for the reflected light to form a spherical wave, one again finds that the Bragg angle is a necessary condition. The locus of the points on each reflecting surface that locate the Bragg angle reflection is traced by the intersection of a sphere through the point light source O_2, and the center of the reflecting system O_1, with the reflecting surfaces. That is, a diametrical cross-section will look like Fig. 7.6b. The resulting virtual source again falls on the sphere in the plane defined by O_1 and O_2. Since the intersection of a sphere with another sphere is a circle, these loci form circular arcs. Again, only light reflected from these arcs contributes to the virtual source. All other reflections contribute undesirable interference.

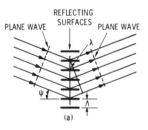

Fig. 7.6. The upper diagram (a) illustrates Bragg reflection from a family of flat reflecting surfaces. The lower diagram (b) represents Bragg reflection from a family of cylindrical surfaces. It also represents a cross section through a diameter of Bragg reflection from a family of spherical surfaces.

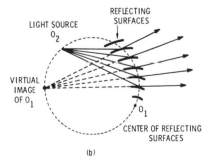

Up to this point we have always chosen the positive sign for the Bragg angle formula of Eq. (7.1). If we choose the negative sign the result is the formation of a reflected wave front converging to a real image. Another point to note is that we have been considering the case for $n = 1$. If the length of the reflecting surfaces is great enough, Bragg angles for $n > 1$ may also be possible, resulting in multiple images which further interfere with the first-order image. Therefore, the interaction region must be kept small.

7.4.2. Bragg Imaging

Since we have arrived at the conclusion that a family of concentric spherical reflecting surfaces will image a point on reflection of a point light source, we can now propose an imaging system. Suppose we provide a family of reflecting surfaces which are made up of the sum of a number of families of spherical surfaces centered on different points. Then, since super-position holds, we would expect the surfaces to reflect a point light source into a replica of the points upon which the different families are centered. Thus we have produced an image, in three dimensions, of the original distribution. Since the Bragg angle is a function of the ratio λ/Λ, we expect an appropriate scaling of the image which can be calculated (again by simple geometry) to be λ/Λ.

Obviously, we can't go around making complicated reflecting surfaces from models we wish to image. We can, however, do the equivalent by using ultrasound to illuminate an object, and shine a light beam through the resulting sound field. Although we are now relying on diffraction of the light by the index of refraction variations induced by the acoustic field, it has been shown that the effect on the light is similar [21].

Up to the present, only cylindrical systems have been used. Therefore, lines parallel to the cylindrical axis are imaged by shadow casting and, consequently, will not suffer the λ/Λ demagnification while transverse lines will. This results in an astigmatic image which must be corrected by cylindrical optics. Several examples of image obtained by this method are shown in Fig. 7.7. Bragg imaging with spherical waves has not been reported. If this can be accomplished, and there appears to be no reason why it cannot, stigmatic images should be obtained.

Bragg diffraction imaging is a convenient real-time method for imaging ultrasonic fields over narrow angles. Thus, it is useful only for transmission objects with a small range of spatial frequencies. A wide range of spatial frequencies will produce a wide-angle sound field with the greatest angles produced by the high spatial frequencies. To image these by Bragg diffrac-

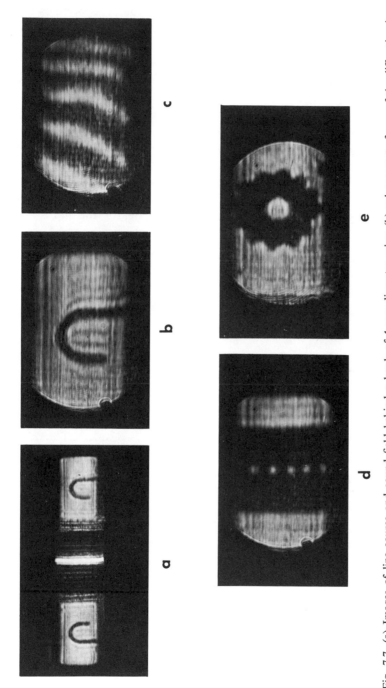

Fig. 7.7. (a) Images of line source and sound field behind a hook of 1-mm-diameter wire; (b) enlargement of one of the diffraction images; (c) Fizeau sound fringes behind a plate almost parallel to transducer; (d) plate with 1.25-mm-diameter holes; (e) small gear as produced by Bragg diffraction. (These photographs are due to A. Korpel and are printed by permission of the author and *App. Physics Letts.* [²¹].)

tion, the field must be probed over a large cross section. This presents the possibility of higher-order image generation, although by keeping the acoustic power low these can be minimized [20]. Nevertheless, the necessity for the illumination of a wide angle of the sound field will result in an increased noise background resulting from light diffracted at other than the Bragg angle.

The primary application for the Bragg technique, we feel, lies in the gigahertz regime. Here, the objects, due to the extreme attenuation of sound at these frequencies, must be very small and the propagating medium must be something other than water; rutile, sapphire, and quartz have been reported in the literature [22,23]. At these frequencies, the ratio λ/Λ nears unity with the result that very high resolutions are achievable. Bragg diffraction imaging, then, may be the prime candidate for the ultrasonic microscope.

7.5. THERMOPLASTIC FILM

The type of detector described here is a substrate coated with a deformable plastic film [24]. The film is heated so that it is soft and then placed in

Table 7.3. Relative Properties of Acoustical Hologram Detectors

Detection method	Sensitivity, W/cm²	Resolution capability	Image formation time, sec
Ultrasound cameras			$10T^b$
1. Quartz	10^{-7}	$3\lambda_w{}^a$	
2. Barium titanate	10^{-9}	$2\lambda_w$	
3. Lead zirconate titanate	10^{-11}	$2\lambda_w$	
Pohlman cell	10^{-7}	λ_w	>1
Liquid surface	10^{-3}	λ_w	$10T$
Scanned single receiver	10^{-12}	λ_w	>30
Receiver array	10^{-12}	λ_w	$10T$
Photographic techniques			
1. Exposure during development	0.05	1 mm	>30
2. Exposure during fixing	1.0	1 mm	>30
Starch plate in iodine	0.5	1 mm	>60

^{a} λ_w = acoustical wavelength in water.
^{b} T = acoustical period in seconds.

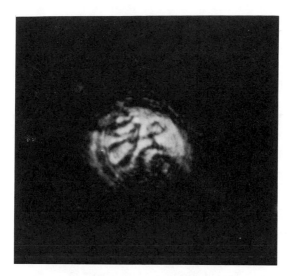

Fig. 7.8. The image from the hologram of script letters GE made on deformable plastic film. (This photograph is due to J. D. Young and J. E. Wolfe and is reproduced by permission of the authors and *App. Physics Letts.* [24].)

the region of interference of the object and reference beams of a holographic system. The film is deformed in relation to the local acoustic pressure and then cooled to form a permanent phase hologram. This method is no different than the liquid-surface system except that a permanent copy of the ripple pattern is obtained. The acoustic intensity required is somewhat higher, as would be expected, due to the higher viscosity required of the detecting film. An example of the reconstruction from such a hologram is shown in Fig. 7.8.

7.6. OPTICAL AND ELECTRONIC READOUT METHODS

In this section we describe a number of techniques for reading the acoustic information without the need of an electronic or acoustic reference.

7.6.1. Optical Heterodyne Technique

The complete description of this technique appears in an article by Massey [25]. In this system a scanning laser beam is used to measure the

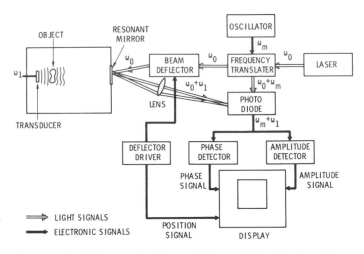

Fig. 7.9. Block diagram for the optical heterodyne technique for recording an acoustical hologram.

vibrations of a resonant glass plate exposed to the acoustic field. Both phase
and amplitude of the vibrations are measured by an optical heterodyne
technique. We consider the system shown in Fig. 7.9. The light from the
laser is passed through a Bragg cell which produces two beams, one at the
original frequency ω_0 and the other at the sum of the original and the
modulation frequency $\omega_0 + \omega_m$, emerging at different angles. One of these
is diverted to be used as a local oscillator and the other is directed through
a scanning system and onto the resonant disk. The disk, being stimulated
by the acoustic field, is vibrating with spatially varying phase and amplitude
at the acoustical frequency ω_1.

The light beam reflected by a point (x, y) on the disk is therefore
phase-modulated due to the changing path length. The amount of phase
shift is expressed as

$$\phi(t) = \frac{4\pi}{\lambda} Z(x, y) \sin[\omega_1 t + \phi_1(x, y)] \tag{7.2}$$

where λ is the wavelength of the light and $Z(x, y)$, $\phi_1(x, y)$ are the spatial
amplitude and phase distribution of the disk deflection. When the reflected
beam is combined with the local oscillator beam and square-law detected,
we end up with the electronic signal

$$i(t) = I_o + I_s + 2\sqrt{I_o I_s} \cos[\omega_m t + \phi(t)] \tag{7.3}$$

which can be rewritten as

$$i(t) = I_o + I_s + 2\sqrt{I_o I_s} \left\{ J_0 \left(\frac{4\pi Z}{\lambda} \right) \cos \omega_m t \right.$$

$$+ \sum_{n=1}^{\infty} J_n \left(\frac{4\pi Z}{\lambda} \right) \cos[(\omega_m + n\omega_1)t + n\phi(t)]$$

$$\left. - \sum_{n=1}^{\infty} J_n \left(\frac{4\pi Z}{\lambda} \right) \cos[(\omega_m - n\omega_1)t - n\phi(t)] \right\} \qquad (7.4)$$

where I_o, I_s are the currents due to the local oscillator and signal beams, respectively.

Since, in general, the amplitude Z is very small we can use the approximations

$$J_0 \left(\frac{4\pi Z}{\lambda} \right) \cong 1$$

$$J_1 \left(\frac{4\pi Z}{\lambda} \right) \cong \frac{2\pi Z}{\lambda}$$

$$J_n \left(\frac{4\pi Z}{\lambda} \right) \cong 0 \qquad (7.5)$$

to obtain the signal

$$i(t) = I_o + I_s + 2\sqrt{I_o I_s} \left(\cos \omega_m t + \frac{2\pi Z(x, y)}{\lambda} \{\cos[(\omega_m + \omega_1)t \right.$$

$$\left. + \phi(x, y)] - \cos[(\omega_m - \omega_1)t - \phi(x, y)]\} \right) \qquad (7.6)$$

Thus, we see that we have preserved, in the electronic signal, the spatial distribution of amplitude and phase of the acoustic pressure on the diaphragm. This information is impressed on the side-band frequencies $(\omega_m + \omega_1)$ and $(\omega_m - \omega_1)$. Since the beam must be swept across the disk to obtain the two-dimensional distribution of the acoustic field, a definite video bandwidth, dependent upon the desired frame rate and the number of sample points required, must be allowed for in the system. This bandwidth will then determine the amount of noise entering the system, and this in turn specifies the sensitivity of the method. Details of this calculation are available in Ref. 25.

The signal centered on one of the side-bands may be phase and amplitude detected and the two signals used to control an oscilloscope trace, with the amplitude signal controlling intensity and the phase signal deviating the trace from a straight line. Since in the final reconstruction we want the twin images to be separated, the resonant disk should be oriented at some angle with respect to the acoustic propagation vector. This introduces a constant term in $\phi_1(x, y)$. That is

$$\phi_1(x, y) = ax + \phi_1'(x, y)$$

thus giving rise to a spatial carrier.

In this system, then, we have suceeded in obtaining a hologram without using an acoustic or electronic reference. The angle of the pick-off surface has supplied this deficiency. Some of the things to look out for in this system are related to the bandwidth requirements. For example, if a television frame rate of 30/sec and 10^4 sample points are required, we need a bandwidth of 600 kHz. Therefore, ω_m and ω_1 must be so chosen that the separation of the two side-band carriers at $\omega_m + \omega_1$ and $\omega_m - \omega_1$ is sufficient to prevent overlap of the doppler spectra.

Although with this system we obtain rapid readout of the acoustic field, we still require a rapid reconstruction system as has been described previously. The advantage this system has over the liquid-surface system is that there is conceptually no limit on the size of hologram that can be read. A disadvantage is, again, the thin resonant membrane required. In this case, however, the membrane need not support a vacuum as in the case of the ultrasonic camera. A disadvantage is the complexity involved in laser scanning systems. Nevertheless, if a good real-time display system is developed, the laser method will probably be useful.

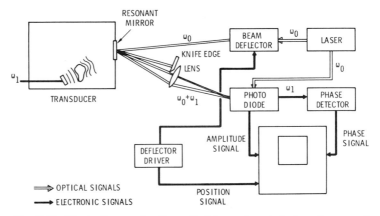

Fig. 7.10. Block diagram for the optical homodyne technique for recording an acoustical hologram. This technique differs from the heterodyne system only in the way in which amplitude information is obtained.

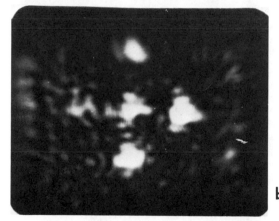

Fig. 7.11. (a) Hologram of five circular sound sources recorded at 9 MHz by the optical homodyne technique; (b) reconstruction; (c) out-of-focus virtual image. (These photographs are reproduced by permission of A. Korpel and P. Desmares and the *J. Acoust. Soc. Am.* [26].)

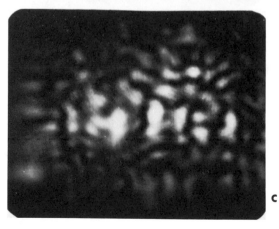

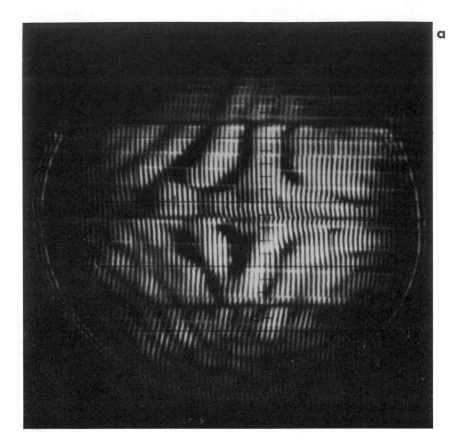

a

7.6.2. Optical Homodyne Technique

This method is quite similar to the optical heterodyne technique except that an optical local oscillator is not used [26]. The explanation of this system is facilitated by Fig. 7.10. The essential difference between this system and the previous one is that the amplitude information is obtained different-ly. Since the vibrations of the surface can be interpreted as traveling waves, the light beam will be swept through an angle depending upon the amplitude of the wave. A knife edge intercepting half the reflected beam when station-ary, has the effect of amplitude modulating the signal when a traveling wave is present, since the beam is oscillated across the knife edge. The phase of the acoustic field is again obtained by phase detection.

This system has the same requirements of bandwidth that was char-acteristic of the optical heterodyne method and hence exhibits the same signal-to-noise ratio and sensitivity limitations. It has the disadvantage that

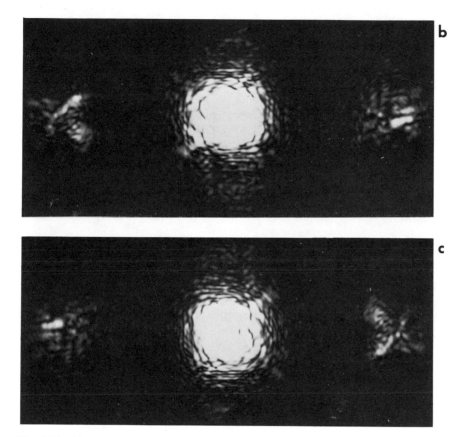

Fig. 7.12. (a) Temporal reference acoustical hologram made at 4.8 MHz with the letter X as object; (b) true image in focus on the left; (c) conjugate image in focus on the right. (These photographs are reproduced by permission of A. F. Metherall and *App. Physics Letts.* [27].)

the knife edge is one-dimensional, whereas the beam motion can be two-dimensional, resulting in some errors. The image offset is again supplied by the canted detection surface.

Examples of holograms and reconstructions obtained by this method are shown in Fig. 7.11.

7.6.3. Temporal Reference

The method described here was proposed and tried experimentally, in an attempt to get around the need for either an acoustic or electronic reference [27]. The proposed method requires the instantaneous measurement and recording of the acoustic field amplitude. For example, the instantaneous complex amplitude of the acoustic field at the point x, y, z at time t is

$$S_1(x, y, z, t) = A_1(x, y, z) \cos[\omega_1 t + \phi_1(x, y, z)] \qquad (7.8)$$

where A_1 and ϕ_1 are the amplitude and phase of the acoustic field, and ω_1 is the acoustic frequency. The theory is that, if this signal can be recorded at a fixed instant of time, say t_h, and t_h is made to advance with one dimension, say x, then the term $\omega_1 t_h$ can take the place of a reference signal. For example, suppose we sweep a sampling beam in the x-direction at a constant velocity, v. Then t_h becomes the linear function of x, x/v and the signal becomes

$$S(x, y, z) = A(x, y, z) \cos\left[\frac{\omega_1}{v} x + \phi_1(x, y, z)\right] \qquad (7.9)$$

Apparently we have supplied an offset and succeeded in measuring amplitude and phase directly. If this signal is added to a bias voltage and used to modulate the intensity of an oscilloscope trace, a photograph of the display will yield a normal hologram.

The practical implementation of this method is again a matter of technique rather than principle. For example, no physical system can make an instantaneous measurement. Therefore, the measurement must be an average over a short time interval on the order of 1/10 the sound period. If the acoustic frequency is 10 MHz this amounts to 10^{-8} sec which requires an electronic bandwidth of 100 MHz. If the averaging time becomes a large fraction of an acoustic cycle, large measurement errors will result.

Since the offset angle is a function of sample sweep rate, v, through the relation

$$2\pi \frac{\sin \theta}{\lambda_1} = \frac{\omega_1}{v} \qquad (7.10)$$

or

$$\sin \theta = \frac{c}{v} \qquad (7.11)$$

the sweep rate for a reasonable angle will be fairly slow. For example, for an angle of 30° and a sound velocity of 1.5×10^5 cm/sec the sweep velocity is 3×10^5 cm/sec. If we require a spatial resolution of one acoustical wavelength, the dwell time or averaging time on each resolution element is λ_1/v, which for a frequency of 10 MHz is 5×10^{-8} sec or one half-period of the ultrasound. For this averaging time one would obviously have considerable measurement error. For this example, then, we could not use the simple Sokolov tube, but would have to switch the beam on for a short interval at each beam position.

The experiment described in Ref. 27 used an acoustical frequency of 4.8. MHz, and a Sokolov tube with a sweep rate of 3×10^5 cm/sec with no switching. Thus, assuming 150 resolution elements across the tube and an electronic beam size of one resolution element, we find the dwell or averaging time to be 0.6 periods of the sound. The resulting holograms cannot be expected to yield sharp images, as seen in Fig. 7.12. Another technique for measuring instantaneous phase by means of double-pulse optical holography is described by Metherell in Ref. 28. However, this technique is also nonreal time so that any advantage this system might have is lost.

REFERENCES

1. P. Greguss, *Ultrasonics Holograms*, Research Film, **5**(4) (1965).
2. M. E. Arkhangel'skii and U. Ia. Afanas'ev, Investigations of the photodiffusion method of visualization of ultrasonic fields, *Soviet Phys-Acoustics* **3**:230 (1957).
3. G. Keck, Acoustical-optical image transformation by means of photographic films, *Acoustics* **9**(2):79 (1959).
4. G. S. Bennett, A new method for the visualization and measurement of ultrasonic fields, *J. Acoust. Soc. Am.* **24**(5):470 (1952).
5. S. Ya. Sokolov, *Means of Indicating Flaws in Material*, U.S. Patent 2,164,125 (1939)
6. P. K. Oschepkov, L. D. Rozenberg, and Ya. B. Semennikov, Electron-acoustic converter for the visualization of acoustic images, *Soviet Phys-Acoustics* **1**:362 (1955).
7. H. W. Jones and H. T. Miles, The development of ultrasonic image converters for underwater viewing and other applications, *Ocean Engineering* **1**:479 (1969).
8. J. E. Jacobs, Performance of the ultrasound microscope, *Materials Evaluation* **25**(3): 41 (1967).
9. H. Boutin, E. Marom, and R. K. Mueller, Real-time display of sound holograms by KDP modulation of coherent light, *J. Acoust. Soc. Am.* **42**:1169 (1967).
10. R. Pohlman, Material illumination by means of acoustic-optical imagery, *Z. Phys.* **113**:697 (1939).
11. R. Pohlman, *Werkwijze en Inrichting voor het Zichtbaar maken van in een Ondorzichtig Medium Aanwezige Insluitsels*, Dutch patent OCT ROOI 48400 (1940).
12. A. Carelli and F. Porreca, Ultrasonic grating remaining after stopping the supersonic wave, *Nuovo Cimento* **9**:90 (1952).
13. A. Carelli and F. Porreca, Ultrasonic grating remaining after stopping the supersonic waves—II, *Nuovo Cimento* **10**:1 (1953).
14. A. Carelli and F. S. Gaeta, Duration of the diffraction grating in relation to the state of the powders in suspension, *Nuovo Cimento* **2**:5 (1955).
15. F. Porreca, On the persistance of a phase grating in some suspensions when stopping the supersonic waves, *Nuovo Cimento* **2**:5 (1955).
16. F. Porreca, On the causes affecting the phase grating permanence at the stopping of the ultrasounds, *Nuovo Cimento* **3**:2 (1956).

17. F. Porreca, Experimental decay law of the diffracted light remaining in the liquids at the stopping of the ultrasonic waves, *Nuovo Cimento* **4**:4 (1956).

18. F. Porreca, On the phase grating in suspensions crossed by ultrasounds, *Nuovo Cimento* **6**:4 (1957).

19. A. Campolattaro, F. Fittipaldi, and F. Porreca, A new method for studying diffusion processes in liquid suspensions by means of ultrasounds, *Nuovo Cimento* **36**:1 (1965).

20. A. Korpel, Visualization of the cross section of a sound beam by Bragg diffraction of light, *App. Phys. Lett.* **9**:12 (1966).

21. A. Korpel, Acoustic imaging by diffracted light. I. Two-dimensional interaction, *IEEE Trans. Sonics & Ultrasonics* **15**:3 (1968).

22. Chen S. Tsai, Harold V. Hance, Optical imaging of the cross section of a microwave acoustic beam in rutile by Bragg diffraction of a laser beam, *J. Opt. Soc. Am.* **42**:6 (1967).

23. J. Havlice, C. F. Quate, and B. Richardson, Visualization of sound beams in quartz and sapphire near 1 GHz, *IEEE Symp. on Sonics and Ultrasonics*, Vancouver, B.C., Canada (1967).

24. J. D. Young and J. E. Wolfe, A new recording technique for acoustic holography, *Appl. Phys. Lett.* **11**:9 (1967).

25. G. A. Massey, An optical heterodyne ultrasonic image converter, *Proc. IEEE* **56**:12 (1968).

26. A. Korpel and P. Demares, Rapid sampling of acoustic holograms by laser scanning techniques, *J. Acoust. Soc. Am.* **45**:4 (1969).

27. A. F. Metherell, Temporal reference holography, *Appl. Phys. Lett.* **13**:10 (1968).

28. A. F. Metherell, S. Spinak, and E. J. Pisa, Temporal reference acoustical holography, Chapter 7, *Acoustical Holography*, Vol. 2, A. F. Metherell and Lewis Larmore (eds.), Plenum Press, New York, 1970.

Chapter 8

Applications

In contrast to optical holography, which has sometimes been described as a solution looking for a problem, acoustical holography was from the outset developed for a specific purpose. This purpose is that of all acoustical probing systems; namely, detecting, locating, and imaging a structure immersed in a medium opaque to electromagnetic radiation. This description encompasses a variety of specific problems, such as nondestructive testing, medical imaging, oil exploration, underwater imaging, etc. In this chapter we describe some of these and include preliminary experimental results where available.

8.1. UNDERWATER VIEWING

The application of acoustical holography to underwater viewing is perhaps the most direct, and at the same time is extremely important. We must bear in mind, however, that the use of holography will not solve the world's problems in underwater viewing. All of the factors working against good acoustic seeing by more conventional means like sonar, also work against the use of holography. We believe, however, that there are certain specialized tasks for which holographic imaging is well suited [¹].

In the preceding chapters we discussed in some detail the various configurations for acoustical holography. It seems obvious that while the liquid-surface method has some advantages over the scanning method at ultrasonic frequencies, these are rapidly neutralized at low kHz frequencies mainly because of the large aperture required for reasonable resolution. Consequently, it seems certain that holographic imaging in the low kHz region will be restricted to the sampling or scanning systems.

8.1.1. Short-Range, High-Resolution Search System

The large variety of modes possible with the scanning systems allows optimum adaptation of the system to the problem. First of all, the generally unstable environment (concerning the scanning platform) requires the fastest possible scanning mode. The candidates would have to be the circular scan and the electronic sampling modes. For short-range imaging (up to 30 m) where it is feasible to use frequencies of 500 kHz to 1 MHz and apertures of 30 to 60 cm, the circular scan mode is probably most economical. Such a system might look like that shown in Fig. 8.1. The array consists of a line of receiving transducers along a radius of a disk. The disk is revolving at a high rate as the boat moves forward. A complete hologram is made for every revolution of the disk. Since this time can be very short (0.01 to 1 sec) it is possible to obtain real time operation.

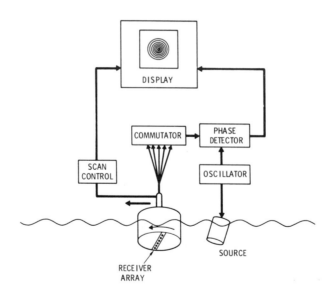

Fig. 8.1. Block diagram for a possible configuration of a high-resolution sonar system utilizing a rotating linear array of receivers.

The actual hologram generation time must be determined with respect to a number of parameters. The velocity of the disk must be slow enough so that bubbles are not generated by the frictional drag. We have found in experiments that this effect is negligible even at rotation speeds of 600 rpm (for a 15-cm-diameter disk). The rotation must be fast enough that platform motion during one rotation is negligible. A good rule of thumb is that the motion should not exceed a quarter of a resolution element in either the lateral or vertical dimension.

As an example let us assume a system operating at 1 MHz with a disk diameter of 60 cm. We wish to have a real-time display so we require 40 frames per second, which requires a disk rotation velocity of 2400 rpm. The angular resolution for this system is $\lambda/A \cong 0.014°$. At a distance of 10 m this results in a horizontal resolution of 2.5 cm and a vertical resolution of 30 cm. Thus, the platform must be constrained to motions of the order of 0.25 m/sec in the horizontal direction. The low horizontal velocity requirement indicated by this example is not practical. This tolerance can easily be relaxed by providing more line arrays on the disk. For instance, four lines of receivers would cut the sampling time by one quarter, thus allowing a 1 m/sec horizontal motion. In addition, one could probably increase the disk velocity.

The system just described might be useful to harbor personnel for location of obstructions, sunken vessels, etc., and to police for location of automobiles, bodies, discarded weapons, and the like. It is also conceivable that the system could be miniaturized sufficiently to allow it to be handled by skin divers to search the bottom in deeper waters. Of course, the only reason for such a system is for operation in waters too silty and murky for optical imaging.

8.1.2. Medium-Range System

By medium range we mean distances on the order of 100 to 500 m corresponding to frequencies in the range of 500 to 100 kHz, respectively. At these wavelengths we need apertures larger than can conveniently be generated by mechanical scanning. Therefore, we consider an electronically scanned array of receivers. If the array is two dimensional it can be instantaneous. That is, each receiver can have its own phase detector that reads and stores the appropriate phase and amplitude at the sampling point. A commutator then reads each value out to the writing and display system. Figure 8.2 is a sketch of such a system. Although we have drawn a rectangular receiver lattice, it is not necessary. One can economize on the number of

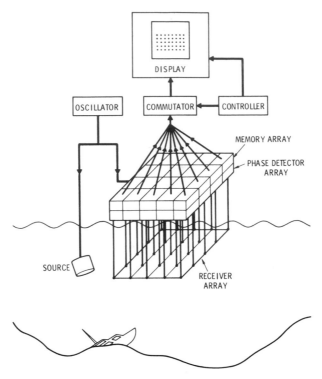

Fig. 8.2. Block diagram for a possible configuration of a high-resolution sonar system utilizing an electronically scanned array of receivers.

receivers by utilizing one of the more generalized sampling patterns as discussed in Chapter 5.

Since the whole array is sampled simultaneously we can relax the stability requirement. The array need remain stationary only for the length of time required to make the measurement, and this is on the order of 10 periods of the sound, or about 20 to 100 μsec. This, of course, assumes that the transmitting transducer remains stationary for the whole time it takes for the sound to travel from it to the target to the receiver. This is an unrealistic requirement for all but fixed systems. However, equivalent operation can be achieved by using pulsed sound and requiring stationarity only over the pulse length. Therefore, a compromise must be struck between energy on target and platform stability. The greater the desired energy in the signal (and correspondingly greater range) the longer the pulse length, resulting in a correspondingly longer stability time. Once the pulse is completed, motion can be resumed since the target will now be insonified and act as a self-luminous source. The stability then required is that of the receiving array as discussed earlier.

Once pulsed operation is introduced, we must consider the possibility of range ambiguity. That is, it is now possible for return echoes from different ranges to arrive at the receiver at the same time. The result of making a hologram with this kind of information is an image consisting of overlapping images of terrain at different ranges. This problem can be overcome by using a sufficiently low pulse-repetition rate. For example, if the maximum range of the system is 300 m we use a pulse interval $T \geq 300/c$, or 0.2 sec. Energy returning from the 300 to 600 m range can then be rejected by the gate set from 0 to 0.2 sec.

This, of course, limits the velocity at which the platform may move without missing parts of the terrain. For example, if the angular coverage of the system is $90°$, the area covered at 300 m is approximately 450 m. Therefore, for complete coverage at 300 m the platform is constrained to move at a velocity $v \leq 2250$ m/sec. This is a rather large velocity for waterborne vehicles and thus presents no problem. However, if ranges from, say, 30 to 300 m are to be included we must use the smaller figure, with the result $v \leq 225$ m/sec. This is still well beyond the capabilities of water craft. However, we see that for longer-range systems where the pulse interval must be lower, range ambiguity can become a serious limitation.

For certain special circumstances it is desirable to use a one-dimensional physical array combined with motion in the orthogonal direction to synthetically generate a two-dimensional array. This can only be done with an exceptionally stable platform since the time to generate the array will be considerable.

In this configuration, the source and receiver are both moving in the direction orthogonal to the array. Therefore, we have simultaneous source and receiver scanning in one dimension but not in the other. Figure 8.3 is an illustration of this system. Equations (4.47), (4.49), and (4.51) show that this kind of scanning results in an astigmatic image. To illustrate, we repeat the necessary equations here while neglecting transit time effects.

$$\frac{1}{r_{b1}} = \pm \frac{k_1}{k_2} \left(\frac{1}{m_1^2 R_0} + \frac{1}{m_4^2 r_1} - \frac{1}{m_4^2 r_2} \right) - \frac{1}{r_a} \qquad (8.1)$$

$$\frac{1}{r_{b1}} = \pm \frac{k_1}{k_2} \left(\frac{1}{m_2^2 R_0} + \frac{1}{m_5^2 r_1} - \frac{1}{m_5^2 r_2} \right) - \frac{1}{r_a} \qquad (8.2)$$

where $1/m_1 = v_\varepsilon/v_\lambda'$, $1/m_2 = v_\mu/v_y'$, $1/m_4 = v_x/v_x'$, and $1/m_5 = v_y/v_y'$. We

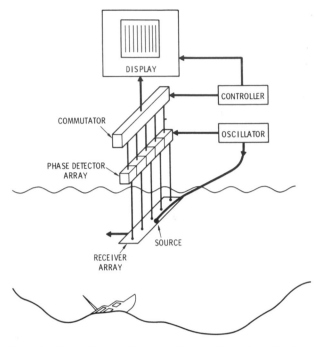

Fig. 8.3. Block diagram for a possible configuration of a high-resolution sonar system utilizing a physically scanned linear array of receivers.

rewrite these equations to more clearly fit our system. First of all we have $R_0 = r_1$. Then we also realize that the source motion is such that the component of velocity v_ε is zero. If we treat the special case of collimated reference and reconstruction beams we have

$$\frac{1}{r_{b1}} = \pm \frac{k_1}{k_2 m_4{}^2} \frac{1}{r_1} \tag{8.3}$$

$$\frac{1}{r_{b2}} = \pm \frac{k_1}{k_2} \left(\frac{1}{m_2{}^2} + \frac{1}{m_5{}^2} \right) \frac{1}{r_1} \tag{8.4}$$

To obtain stigmatic imaging, these two values for r_b must be equal, with the result

$$\frac{1}{m_4{}^2} = \frac{1}{m_2{}^2} + \frac{1}{m_5{}^2} \tag{8.5}$$

or

$$\frac{1}{m_4{}^2} - \frac{1}{m_5{}^2} = \frac{1}{m_2{}^2} \tag{8.6}$$

That is, the astigmatism generated by unequal scanning ratios in the two dimensions can be compensated by unequal recording of the hologram in the two dimensions. An alternative procedure is to record the hologram without dimensional distortion and then use an astigmatic optical system to restore the image.

We present a numerical example to better illustrate the problems and the solutions. We assume the hologram is made at a frequency of 250 kHz. A region 2 m square is swept out by the array. The enlargement ratio between the acoustic aperture and the hologram, $1/m_2$, is assumed to be 20 so that $v_y' = v_\mu/20$. Using Eq. (8.6) we have $1/m_4^2 - 1/m_5^2 = 400$, or $(v_x/v_x')^2 - (20v_y/v_\mu)^2 = 400$. Since we stated earlier that $v_\mu = v_y$ we have, finally, $(v_x/v_x') = (800)^{\frac{1}{2}} \cong 29.3$. Thus, our hologram will be 7.1 cm in the x-direction and 10 cm in the y-direction. Figure 8.4 may help to understand this example.

If we choose to correct the astigmatism in the reconstruction step we set $m_4 = m_5$ and solve for the two focal distances. Then we have

$$r_{b1} = c_1 r_1 \qquad (8.7)$$

$$r_{b2} = c_2 r_1 \qquad (8.8)$$

where

$$\frac{1}{c_1} = \frac{k_1}{m_4^2 k_2}, \qquad \frac{1}{c_2} = \frac{k_1}{k_2}\left(\frac{1}{m_2^2} + \frac{1}{m_4^2}\right)$$

If we use a point object, this means that the hologram will reconstruct two line images orthogonal to each other at the distances prescribed by Eqs.

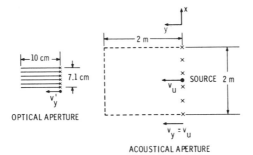

Fig. 8.4. A numerical example of an acoustical hologram made by sweeping out an aperture with a line array of receivers and a source. The recorded aperture is distorted to compensate for astigmatism introduced by this method.

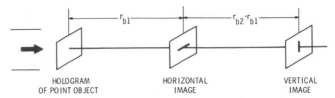

Fig. 8.5. This diagram illustrates the astigmatism introduced by the system shown in Fig. 8.4 when the recording is not distorted. The object in this illustration is a point.

(8.7) and (8.8) as shown in Fig. 8.5. To form a single point image we must insert a cylindrical lens somewhere in the system in such a way as to accomplish this.

Since we can think of the hologram as an astigmatic lens, it is only necessary to place a cylindrical lens next to the hologram so that the lens–hologram combination becomes stigmatic. The example shown in Fig. 8.5 could be corrected by a convex cylindrical lens of focal length determined by the simple lens formula $1/f + 1/r_{b2} = 1/r_{b1}$, placed so its power is exerted in the horizontal dimension. Alternatively, a concave cylindrical lens could be placed with its power in the vertical dimension so as to move r_{b1} out to r_{b2}. A third alternative is to place the lens between r_{b1} and r_{b2}. In this case, since we are working with the images rather than the hologram, we must take into account the fact that the magnifications are different by the factor 2 as described in Chapter 4. That is, the line image at r_{b1} is twice as long as the line image at r_{b2}. Thus, we have the simultaneous equations

$$\frac{1}{f} = \frac{1}{D_1} + \frac{1}{D_2} \tag{8.9}$$

$$M = \frac{D_2}{D_1} = \frac{1}{2} \tag{8.10}$$

where $D_1 + D_2 = r_{b2} - r_{b1}$ and D_1 is the distance from r_{b1} to the lens. Solving for D_1 we obtain

$$D_1 = \tfrac{2}{3}(r_{b2} - r_{b1}) \tag{8.11}$$

For the particular example we are using, $(v_\mu = v_y, v_x/v_x' = v_y/v_y')$, we have $r_{b2} = 2r_{b1} = 2c_1 r_1$ and $D_1 = \tfrac{2}{3} c_1 r_1$. Figure 8.6 shows an example of a hologram reconstructed in this way with $k_1/k_2 = 250$, $m_4 = m_5 = 3$, $f = 20$ cm, and $r_1 = 30$ cm.

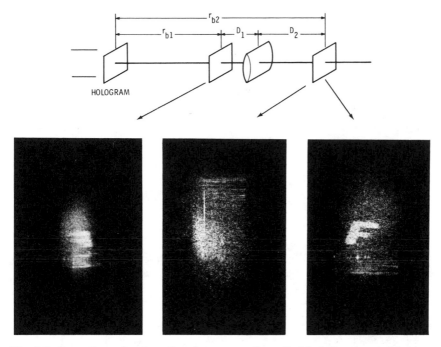

Fig. 8.6. Correction of astigmatism by means of a cylindrical lens inserted between horizontal and vertical focus. The object consisted of the letter F and a ball bearing. The ball bearing was used so that horizontal and vertical focus could be easily located. The photographs, from left to right, are taken at the plane of horizontal focus, the plane of vertical focus without the lens, and the plane of vertical focus with the lens.

We note that the optical methods of restoring the image depend on the object distance r_1. Therefore, this method is not as general as the distorted recording method.

An interesting application based on the reciprocal of the scanned line array is one in which both source and line array are stationary and the object moves through the field. This system is exactly equivalent to the last one, and requires the same compensated recording. The obvious application for such a system might be in continuous harbor entrance surveillance as shown in Fig. 8.7. The submarine provides the scanning action required to build up a synthetic aperture.

In all of these systems, although we have drawn the diagrams showing a single point source, multiple sources are acceptable, indeed desirable. This is because specular targets may be missed entirely if illuminated from one direction only. Multiple sources tend to simulate diffuse illumination which results in much better images. We must remember, however, that the

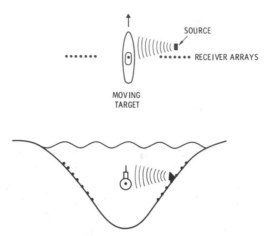

Fig. 8.7. Conceptual design for a channel surveillance system utilizing fixed receiver and source arrays. The transiting vehicle generates its own hologram, as predicted by the reciprocity principle.

sources must be rigidly coupled so that the phase front remains the same over the recording time. That is, the source remains spatially coherent.

Figure 8.8 is a demonstration of the improvement in the image of a specular target when diffuse illumination is used.

8.1.3. Long-Range Systems

It is doubtful that holography has anything to offer for long-range acoustical imaging. The reasons for this are the usual ones of high attenuation and aberrations induced by the medium. Since the resolution at long ranges is so poor that targets generally appear only as points, it is more efficient to use the well-developed pulse-compression techniques along with beam forming systems [2].

8.2. GEOPHYSICAL APPLICATIONS

The techniques used in geophysical exploration are very similar to those used in undersea viewing. The problem is considerably complicated by the increased complexity of the medium and the extreme attenuation of the sound waves. For this reason the sound source is generally a high-explosive charge that generates a wide spectrum of frequencies of which

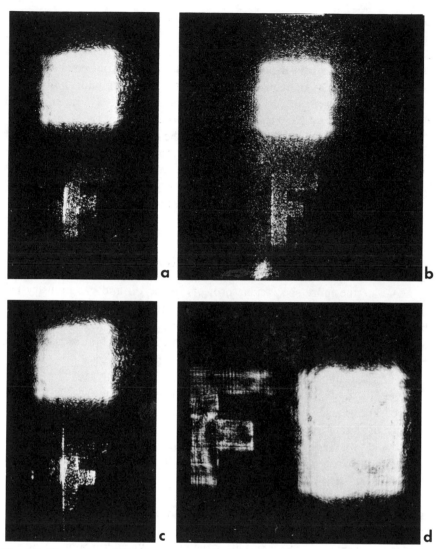

Fig. 8.8. Illustration of the importance of diffuse illumination for specular targets. Photographs (a), (b), and (c) are images from acoustical holograms of a specular target illuminated by a spherical wave, diffuse wave obtained by reflection from a piece of styrofoam, and a scanning point source, respectively. The fourth photograph (d) was obtained from a hologram made by the simultaneous source–receiver scan method. Note that some diffuse illumination was present in (a) and (c), probably due to reflections from other structures in the tank. Due to the finite receiving cone, it is difficult to obtain complete coverage of the object even though we know where it is. The simultaneous source–receiver scan completely eliminates this problem, especially if the same transducer is used for both.

only the lower ones (1 to 60 Hz) are routinely used. Nevertheless, acoustical holography has been proposed for use in geophysical exploration [4,5]. Since coherent radiation is required, one of the coherent sources (such as the Vibro-seis thumper) would need to be used [6]. If powerful, compact coherent radiators at frequencies of up to 200 Hz can be built, holography might become practical for this application. Alternatively, narrow band filters can be used to extract a coherent component from the signal.

We will not attempt to explore in detail the feasibility of holography for geophysical exploration. This is best left to oil exploration scientists thoroughly familiar with the acoustical problems. The basic theory of holography as developed in earlier chapters is unchanged for this application. We will, however, discuss some special problems and techniques which may be useful.

8.2.1. Subsurface Cavity Mapping

One of the more spectacular projects in the oil and gas industry has been the one involving atomic explosives to create a cavity in oil or gas bearing strata. The idea is to stimulate oil or gas flow into the large cavity created by the blast, drilling a borehole into the cavity and extracting the product. This technique could assume great importance since large reserves are trapped in impermeable shales and in tar sands.

One of the problems in the evaluation of this technique has been the lack of knowledge of the size and shape of these cavities. The present method of evaluation is simply to drill holes over an area, hoping to intercept the cavity. This is both expensive and inaccurate. We propose here a use for holography which was first demonstrated in the optical literature [7], and later extended to the acoustical domain in our laboratories [8].

The basic idea involved here is that of differential holography. We arrange to make a double exposure hologram of the subsurface in the area of the proposed cavity as shown in Fig. 8.9. The first hologram is made before the event and the second after the event. We can arrange to display the two holograms in such a way that the reconstructed image consists of the change occuring between the two exposures. Then, only the cavity will be imaged. We now proceed to prove the feasibility of this scheme.

For simplicity, we assume the volume to be made up of a collection of point scatterers. We further assume that, if we illuminate such a volume, each point will scatter energy back at the detector without appreciable interference from the other particles. This is perhaps an unrealistic model, but it presents the advantage of being workable. Thus, the complex amplitude at

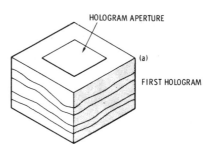

HOLOGRAM APERTURE

(a)

FIRST HOLOGRAM

Fig. 8.9. Conceptual portrayal of a cavity-measurement system for use in atomic explosive experiments. The two holograms are made in such a way that together they image the cavity only, thus eliminating extraneous information generated by the complicated geological features.

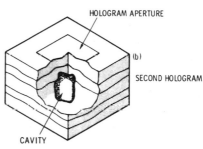

HOLOGRAM APERTURE

(b)

SECOND HOLOGRAM

CAVITY

a point in the hologram plane due to the collection of scatterers is

$$A_1(x) = \sum_{n=1}^{N} a_n \exp(i\phi_n) \exp(i\omega t) \tag{8.12}$$

where ω is the acoustic frequency. We assume that the explosion removes a certain number of the scatterers, so that the signal at the same point x, after the event is

$$A_2(x) = \left[\sum_{n=1}^{N} a_n \exp(i\phi_n) - \sum_{n=l}^{l+m} a_n \exp(i\phi_n) \right] \exp(i\omega t) \tag{8.13}$$

where $m < N$ particles have been removed. As we know from previous chapters, these complex signals can be recorded by phase detection and stored on magnetic tape film. Suppose we choose to store it for later use. We phase detect electronically by mixing the signal with the reference signal $\exp i(\omega t + \alpha x)$ to obtain

$$S_1(x) = \left\{ \sum_{n=1}^{N} a_n \exp[i(\phi_n + \alpha x)] \right\} \tag{8.14}$$

and

$$S_2(x) = \left\{ \sum_{n=1}^{N} a_n \exp[i(\phi_n + \alpha x)] - \sum_{n=l}^{l+m} a_n \exp[i(\phi_n + \alpha x)] \right\} \tag{8.15}$$

There are a number of ways in which this information can be recorded on a single hologram. The simplest and best way is to electronically subtract them. If this is done we obtain the signal

$$S_3(x) = \left\{ \sum_{n=l}^{l+m} a_n \exp[i(\phi_n + \alpha x)] \right\} \tag{8.16}$$

which is recorded as

$$\sum_{n=l}^{l+m} a_n \cos(\phi_n + \alpha x) \tag{8.17}$$

which we see contains information about the deleted particles only. In other words, the information about the unchanged part of the volume has been removed.

The signal $S_3(x)$ is used to modulate a light spot as it is scanned across a piece of photographic film. To be able to record negative quantities we use a bias voltage and modulate the light intensity about an average intensity. Therefore, the transmission function of the hologram becomes

$$T = K\left[a_0 + \sum_{n=l}^{l+m} a_n \cos(\phi_n + \alpha x) \right] \tag{8.18}$$

If we now illuminate this hologram with light, we obtain an image of the cavity. The twin images are, of course, separated due to the use of the phase shift term αx.

We have simulated this experiment in the laboratory. In this particular experiment we did not have equipment available for magnetic recording of the signals. Therefore, we performed our subtraction differently. The first signal was recorded on film by mixing with the reference signal $\{\exp[i(\omega t + \alpha x)]\}$ and the second by mixing with $\{\exp[i(\omega t + \alpha x + \pi)]\}$ with the result

$$S_1(x) = \sum_{n=1}^{N} a_n \exp[i(\phi_n + \alpha x)] \tag{8.19}$$

$$S_2(x) = \left\{ -\sum_{n=1}^{N} a_n \exp[i(\phi_n + \alpha x)] + \sum_{n=l}^{l+m} a_n \exp[i(\phi_n + \alpha x)] \right\} \tag{8.20}$$

These signals were written on film with suitable bias voltages to form the holograms

$$T_1 = K\left\{ a_0 + \sum_{n=1}^{N} a_n \cos(\phi_n + \alpha x) \right\} \tag{8.21}$$

$$T_2 = K\left\{ a_0 - \sum_{n=1}^{N} a_n \cos(\phi_n + \alpha x) + \sum_{n=l}^{l+m} a_n \cos(\phi_n + \alpha x) \right\} \tag{8.22}$$

In the experiment, both holograms were recorded on the same piece of film by interlacing the scan lines. When this hologram is reconstructed the resulting wave front will contain only the information expressed in the terms

$$a_0 + \sum_{n=l}^{l+m} a_n \cos(\phi_n + \alpha x) \qquad (8.23)$$

The experimental proof of these statements is shown in Fig. 8.10.

The importance of this technique lies in the fact that extraneous information that may confuse the issue is eliminated. Thus, all the sound waves (suffering reflections and refractions in the various sedimentary layers) not contributing to the image are not present. This, of course, does not exempt the desired information from distortion for the same reasons. The method merely removes clutter. We next describe an additional technique that may be useful in removing the distortion from the remaining image.

8.2.2. Distortion Correction

The method described here was first developed in the optical domain [9], and involves the making of a hologram of the distorting medium itself. This hologram is then used to correct the distortion imposed on the image of the cavity by the medium.

Let us assume that the two signals discussed in Section 8.2.1 are written as

$$S_1(x) = \sum_{n=1}^{N} a_n \exp[i(\phi_n + \varepsilon_n)] \exp(i\omega t) \qquad (8.24)$$

and

$$S_2(x) = \left\{ \sum_{n=1}^{N} a_n \exp[i(\phi_n + \varepsilon_n)] - \sum_{n=l}^{l+m} a_n \exp[i(\phi_n + \varepsilon_n)] \exp(i\omega t) \right\} \qquad (8.25)$$

where ε_n is the error introduced by the medium. We further assume that we record these signals, subtract them and write the result on film to produce the hologram designated by

$$T_1(x) = K \left\{ a_0 + \sum_{n=l}^{l+m} a_n \cos(\phi_n + \varepsilon_n + \alpha x) \right\} \qquad (8.26)$$

Now suppose that before the explosive is set off, we make a third hologram. This one, however, is made by lowering an acoustic source of the same frequency into the borehole in which the explosion is to be set off. We then make a hologram of the source as distorted by the medium. The

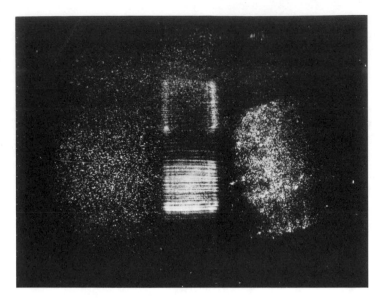

a

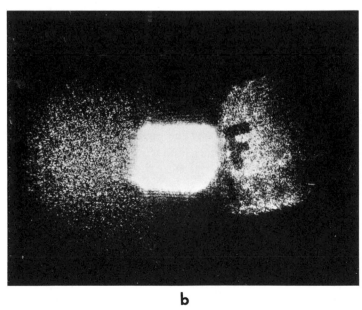

b

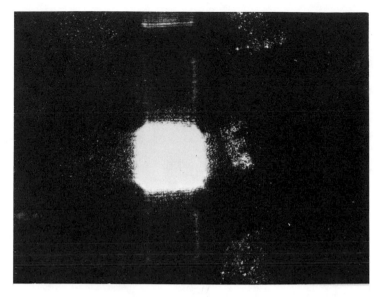

c

Fig. 8.10. The experimental verification of the concept described in Fig. 8.9. The top photograph (a) is an image from an acoustical hologram of a styrofoam sheet, (b) is an image of the same sheet after part of it has been removed, and (c) shows the image of a hologram made by interlacing two holograms. The interlaced hologram consisted of one using the sheet as object and one using the sheet with the F removed as object and a π radian phase shift in the reference. The resulting reconstruction subtracts the common area of the two images, leaving only the removed area, namely, the F.

signal at the surface will be

$$S_3(x) = a_3 \exp[i(\omega t + \varepsilon)] \qquad (8.27)$$

This signal is recorded as the hologram

$$T_2(x) = K[a_0 + a_3 \cos(\varepsilon + \alpha x)] \qquad (8.28)$$

The two holograms, expressed in Eqs. (8.26) and (8.28) are then used in the optical system shown in Fig. 8.11. Therefore, the correction hologram is illuminated with the light field expressed by

$$b \sum_{n=l}^{l+m} a_n \exp[i(\phi_n + \varepsilon_n + \alpha x + \omega t)] \qquad (8.29)$$

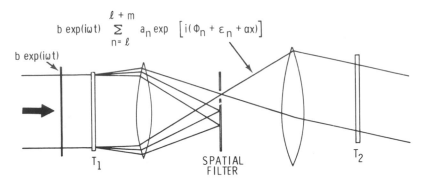

Fig. 8.11. The optical arrangement to perform error correction due to a random intervening medium is shown here. The hologram of the object imbedded in the distorting medium is labeled T_1, while the hologram of a point object embedded in the same medium is labeled T_2. A mask placed in the spatial frequency domain selects the desired image term. This portion of the light illuminates the second hologram which then performs a cancellation of the error terms as described in the text.

with the result that the light wave leaving it is expressed by

$$
Kb\left\{ a_0 \sum_{n=l}^{l+m} a_n \exp[i(\phi_n + \varepsilon_n + \omega t + \alpha x)] \right.
$$

$$
+ a_3 \sum_{n=l}^{l+m} a_n \exp[i(\phi_n + \varepsilon_n + \varepsilon + 2\alpha x + \omega t)]
$$

$$
\left. + a_3 \sum_{n=l}^{l+m} a_n \exp[+i(\phi_n + \varepsilon_n - \varepsilon + \omega t)] \right\} \qquad (8.30)
$$

We see that there are three waves propagating in three clearly defined directions. If the distorting medium is relatively slowly varying, then ε_n will approximate ε with the result that the error terms are canceled to yield an image term

$$
Kba_3 \sum_{n=l}^{l+m} a_n \exp[i(\phi_n + \omega t)] \qquad (8.31)
$$

which is exactly the image term, undistorted.

This technique will work exactly for only one particle, namely that particle occupying the same position as the source for the correction hologram. However, if the distortion is large in scale compared to the size of the cavity, all error terms will be approximately corrected.

In summary, we outline once again the procedure for measuring an explosion-induced cavity in the earth. First, the acoustic source is lowered into a borehole to the position where the explosive is to be set off. A receiver is scanned over an aperture on the surface and the phase and amplitude at each position are measured and used to make a hologram. Then the source is withdrawn and coupled to the surface while another receiver scan is made. This time the received signal is stored on magnetic tape. All equipment is removed and the blast is set off. The equipment is replaced in the exact location used earlier and another scan is made and recorded. The two tapes are then played back in exact synchronism, the signals are subtracted, and the resulting signal used to make the second hologram. The two holograms are then used to obtain a distortion corrected image of the cavity. Alternatively, the first scanned data may be used to correct the subtracted data by computer processing.

It might be advantageous to use simultaneous source and receiver scanning for better resolution. In this case, the distortion correction hologram must be made differently. This could be done by lowering a repeater into the borehole to the explosive site. The source and receiver are then scanned over the surface. The repeater is triggered when the sound from the source reaches it, amplifies the signal, and sends it back to the receiver. This signal must be so large as to overwhelm the reflections returned by the subsurface structure. In this way we obtain a distortion correction hologram which is used to correct the simultaneous source–receiver scan hologram.

8.3. NUCLEAR REACTOR SURVEILLANCE

There are certain devices and machines that are required to operate reliably over long periods of time. In fact, reliability must be very nearly perfect. Furthermore, in some of these devices, breakdown can have catastrophic results. A nuclear power reactor is perhaps the most dramatic example of such a device. In addition, even if the shutdown is not catastrophic, it is extremely difficult to find the cause because of high radiation levels.

In some water-cooled reactors, it is possible to drain the coolant and insert optical periscopes for viewing the interior. The later models of power reactors are expected to be liquid-metal cooled. This means that even if reactor shutdown occurs, the liquid metal will be kept in circulation to avoid the great cost of draining it. Consequently, optical devices are out of the question. Ultrasound, however, is readily propagated in liquid metals. Therefore, we feel that the appropriate application of ultrasonic holography may be useful.

We describe here two possible modes of operation for ultrasonic systems for this application. The first of these might be termed an ultrasonic periscope which is an internal device. The second system would make use of the analogy of viewing the contents of a transparent glass vessel by means of optical holography.

8.3.1. Ultrasonic Holographic Periscope

We begin with the basic assumption that the reactor is shut down, but the liquid-metal coolant is not drained. This assumption is required because during operation the radiation and temperature levels inside the reactor are so high that ultrasonic transducers become inoperative. During shutdown, however, radiation levels are reduced and the temperature of the coolant is maintained just above its melting point (for sodium this is around 600°F). It is then feasible to think of inserting the ultrasonic transmitter and receiver into the coolant.

The system consists of a rotating scan head, mounted at the end of an articulated telescoping arm much as the presently used periscope, and a display system of the type shown in Fig. 8.12. Notice that we have indicated

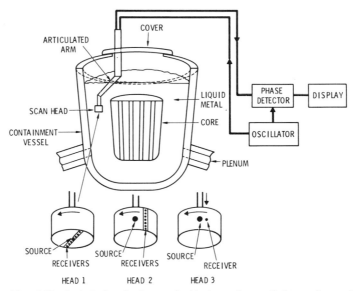

Fig. 8.12. Artist's sketch of acoustical holography applied to an internal periscope for liquid-metal-cooled reactors. The inset shows three possible scanning heads designed for different purposes as described in the text.

three different scan heads. The first head is designed to look down, or if the articulated arm can be placed in the horizontal position it can also look sideways. The other two heads are designed to look all around over a 360° arc. Head 2 has the astigmatic scanning we discussed earlier. Hence, astigmatic recording is required to obtain a stigmatic image. The third head is designed to perform a continuous 360° search. Scanning is accomplished by rotation and translation of the drum. Since both source and receiver scan at the same rate in both dimensions, we obtain the advantages of increased resolution and a larger image.

There are no new techniques peculiar to this application, only novel engineering designs for the scanner. The major problem appears to be one of planning the reactor to allow for inserting the device. Since present water-cooled reactors already allow for the insertion of optical devices this requirement does not seem excessive. The other areas of concern are the development of the display system, and of ultrasonic transducers able to withstand rather hostile environments.

8.3.2. External Surveillance System

The system we have in mind here might be termed "way out" [10]. If it can be made into a practical system, however, it could play an important part in reactor technology. This particular system is one in which all equipment is external to the containment vessel. Since steel is transparent to sound, the idea is to transmit sound through the wall into the coolant. The sound reflected from the contents is returned through the wall and detected by a scanned receiver coupled to the wall by some liquid medium. A simplified system of this type is shown in Fig. 8.13.

Any number of transducers may be attached to the vessel wall to provide a full coverage sound field inside the reactor. At various strategic positions around the reactor we have areas where the scanning apparatus may be mounted. The rest of the system is no different than any of the others.

The problems likely to arise here are, in order of importance: (1) designs of the reactor and instrumentation to allow for clear areas for the scanner; (2) multiple reflections in the reactor wall; and (3) generation of shear waves at the interfaces. The design of the reactor is listed here as first in importance because, even if a perfect imaging system were available, present reactor designs are so cluttered up with various structures and instruments, and the walls so inaccessible that it is doubtful that space could be found for it. Therefore, if the decision is made that such a viewing system is necessary, the reactors must be designed to allow for clear areas of access.

Once access is provided, technical considerations take over. In laboratory experiments with this system we have found that multiple reflections in the wall are not negligible. However, for the most part these can be discriminated against on a time basis by using pulsed ultrasound together with time gating at the phase detector. The other technical problem is the possibility that shear waves generated at the interfaces may cause interference effects. Here, again, since shear waves travel slower in solids than do longitudinal waves (which are the ones normally used since they propagate in liquids while shear waves do not) they can perhaps be discriminated against by time gating.

Both of the problems mentioned above are not connected with holography *per se*; they involve the problem of delivering coherent acoustical energy to the medium and back to the receiver. Since the laws of optics hold for sound, with the exception of shear-wave generation, many of the solutions derived in the optical domain may be applicable. For example, when imaging through glass plates with light, we make use of high-quality glass and antireflection coatings. It is entirely possible to do the same in the acoustical domain. This may necessitate special inserts in the reactor vessel which might be dubbed high-quality "acoustical windows."

The generation of shear waves may also be reduced by designing an acoustical window with special curved surfaces so that the sound waves

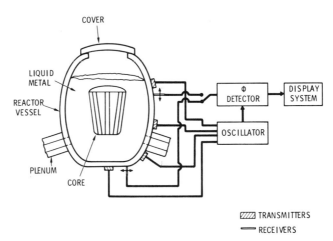

Fig. 8.13. Artist's sketch of acoustical holography applied to an external surveillance system for liquid-metal-cooled reactors. The interior of the reactor vessel is flooded with ultrasound from sources stationed around the vessel. At selected areas of the reactor are scanning systems to collect scattered sound to form a hologram.

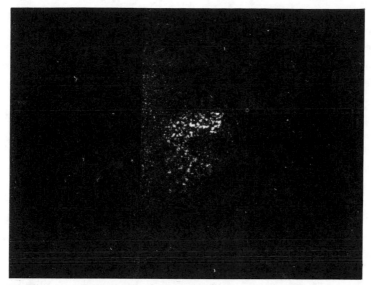

Fig. 8.14. This photograph shows the image from a hologram made
through a 2-in.-thick aluminum plate.

reflected from the objects tend to strike the window normal to its surface.
Shear waves are generated when longitudinal waves strike a solid interface
with amplitudes in direct proportion to the angle of incidence. Thus, if the
longitudinal waves arrive normal to the surface, no shear waves are
generated.

Some laboratory results have been obtained with a 2-in.-thick aluminum
barrier and water as the medium. These are shown in Fig. 8.14 along with
details of the experiment.

8.4. MEDICAL IMAGING

Ultrasonics has been used in the field of medicine since 1937 when the
Dussik brothers in Austria attempted to image the brain by measuring the
attenuation of an ultrasonic beam transmitted through the head [11]. Since
that time, considerable progress has been made using, for the most part,
echo location techniques. A narrow, pulsed ultrasonic beam is swept, either
by hand or automatically, over a line on the body surface. The return echoes
are made to modulate a cathode-ray spot that is controlled by motion sen-
sors on the transducer in such a way that the resulting display portrays a

cross section of the body. Very good images have been obtained of the soft tissue portions of the body.

This technique, when applied to areas of the body having a bone covering (such as the head or chest area) have not been successful because of the interference induced by the intervening bone structure. The interfering effects arise because of the very high attenuation of sound in bone, its scattering properties, and the multiple reflections between bone boundaries. The last effect is most severe in the case of imaging the brain. The whole field of ultrasonic echoencephalography is well described by D. N. White, and will not be elaborated on in this book [12].

Since the basic technique of holography is no different in this application than in any other, we will only enumerate the advantages we foresee in this application. We recognize, of course, that the problems encountered in conventional echoencephalography will still be problems for holography. We believe, however, that at least some of them will be alleviated.

Attenuation will always be a problem. With holography, however, we can use a multitude of sources located around the head or body to provide relatively high acoustical fields without applying excessive power to any one point on the surface. An added advantage is that this produces a diffuse illumination, thus eliminating the problem of specular reflections from planar surfaces.

Resolution will be improved since we can provide a much larger aperture than that used with present techniques. Presently, a single transducer, used as both transmitter and receiver, provides a fine beam with resolution dependent upon the beam diameter. The size of the transducer is limited since it must be used to scan a line. With holography, we use a very small receiver to generate a large synthetic aperture with its attendant fine resolution.

The problems associated with the bone covering are also somewhat alleviated when holography is used. In particular, since many low-powered sources can be used in place of one high-powered source, the problem of multiple reflections between the bone boundaries is much reduced. Any remaining vestige of multiple reflections can easily be discriminated against by time gating at the receiver. The one effect remaining is distortion of the return energy, and hence the image, due to irregularities in the bone covering. Unfortunately, this is unavoidable. However, having reduced the other problems, we feel that superior imaging, even through bone covering, can be achieved.

Since different parts of the body present different instrumentation problems we can conveniently discuss them in terms of scanning or liquid-surface holography.

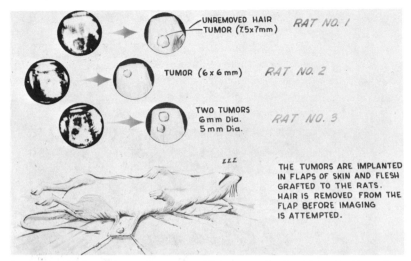

Fig. 8.15. This figure shows the way in which tumors implanted in skin flaps of rats were imaged by acoustical holography. The tumors varied in size from 5 to 7.5 mm, the frequency was 9 MHz, and the detection method was liquid-surface levitation.

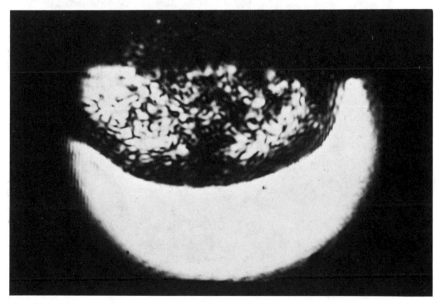

Fig. 8.16. One of the potential areas of application for liquid-surface acoustical holography is mammography. The picture shown here is the image at 3 MHz of a normal breast of a 54-year-old woman. The reticulated pattern represents breast structure. Tumors may be expected to produce dark areas with structure and characteristics which can be identified as different from normal structure.

8.4.1. Liquid-Surface Holography

In general, liquid-surface holography is most applicable where trans-illumination is possible. This, then, includes most appendages of the human body. The female breast is particularly amenable to this technique since it is readily accessible and made up entirely of soft tissue. This very charac-

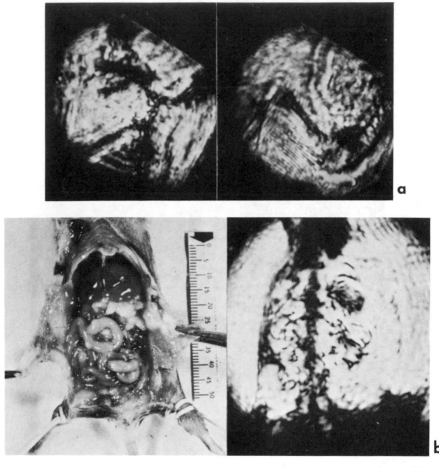

Fig. 8.17. This is a collection of images obtained with the liquid-surface system to illustrate the potentialities of acoustical holography for medical applications. (a) Three-month-old human fetus. Bone development is readily observed in these reconstructed holograms. Note the fingers, cervical discs, ribs, lumbar discs, and knee cartilage. Skull development shows mouth/sinus cavity, eye and ear location. (b) Mouse with cancer. The light-colored nodules between the liver and the intestinal loop are cancer. In the reconstructed hologram the cancer can be seen as a dark area.

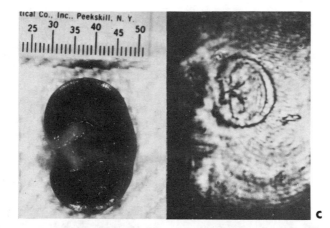

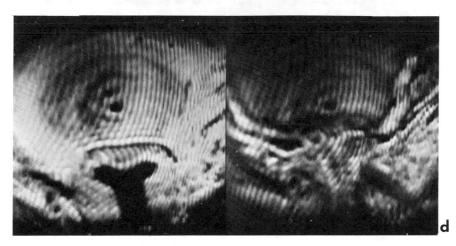

Fig. 8.17 (*continued*). (c) Healthy rabbit kidney. The calyx, medullar, and cortex region show well in both the photograph and the reconstructed hologram. (d) 1, Human uterus. 2, Human uterus with air in lumen. The hologram of this excised human uterus shows the ovaries, tubes, and lumen. Note the improvement in lumen viewing when air is introduced as a constrasting method.

teristic makes X-rays difficult to read since there is so little variation in density.

Preliminary experiments have shown that tumors are much more opaque to ultrasound than normal body tissue. Figure 8.15 shows the results of tests on live rats with tumors implanted in skin flaps. Further evaluation performed at Roswell Park Memorial Hospital in Buffalo, New York,

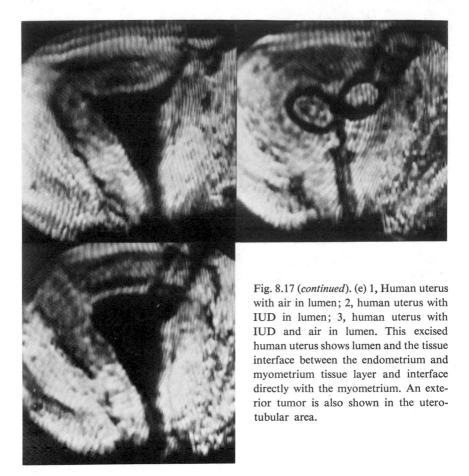

Fig. 8.17 (*continued*). (e) 1, Human uterus with air in lumen; 2, human uterus with IUD in lumen; 3, human uterus with IUD and air in lumen. This excised human uterus shows lumen and the tissue interface between the endometrium and myometrium tissue layer and interface directly with the myometrium. An exterior tumor is also shown in the uterotubular area.

reported that such tumors were much more readily seen by ultrasonic holography than by X-ray radiography [13].

A propotypical system for breast imaging has been built and partially tested with inconclusive results. We are confident, however, that further work will show that this application will be useful. A typical image obtained with this system is shown in Fig. 8.16. Note what appear to be the lactic ducts converging at the nipple area.

Experimental images obtained of a variety of biological subjects are shown in Fig. 8.17. Remembering that these images are made with prototype equipment and perhaps harkening back to early X-ray pictures, the reader will agree with us that this type of ultrasonic holography may be a useful replacement or adjunct to X-rays, particularly if ionizing radia-

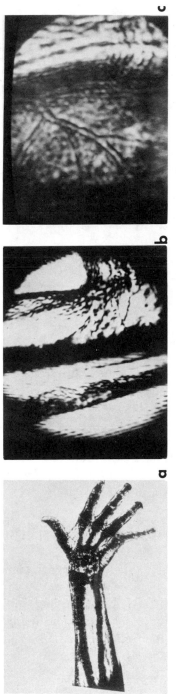

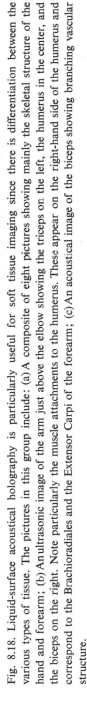

Fig. 8.18. Liquid-surface acoustical holography is particularly useful for soft tissue imaging since there is differentiation between the various types of tissue. The pictures in this group include: (a) A composite of eight pictures showing mainly the skeletal structure of the hand and forearm; (b) An ultrasonic image of the arm just above the elbow showing the triceps on the left, the humerus in the center, and the biceps on the right. Note particularly the muscle attachments to the humerus. These appear on the right-hand side of the humerus and correspond to the Brachioradiales and the Extensor Carpi of the forearm; (c) An acoustical image of the biceps showing branching vascular structure.

Fig. 8.19. This image was obtained with the scanning system. The object consisted of a styrofoam letter B imbedded in the brain of a seven-day-old pig. The letter was about 2 cm below the skull cap which was about 7 mm thick. The hologram was made at 9 MHz with time gating used to reject reflections from the skull.

tion is to be avoided. Images of the hand, forearm, and upper arm produced on improved equipment manufactured by Holosonics, Inc., are shown in Fig. 8.18.

8.4.2. Scanning Holography

The scanning system of holography is most applicable where large attenuation and, hence, low signal levels are to be expected. These conditions generally occur for reflection holography, especially when a bone covering must be penetrated or relatively large distances are involved, such as the abdominal area. The abdominal area, since it is soft tissue, may be amenable to liquid-surface holography also.

An additional factor to be considered is that the relative motion between aperture and object must be small during the scan. This factor alone may limit scanned holography to the imaging of the brain structure, since the abdominal region tends to be in constant motion, both internally and externally.

A preliminary experiment involving the imaging of an artifact implanted in the brain of a pig is shown in Fig. 8.19. Again, although the result may not appear impressive, it is certainly encouraging. Time gating was used to accept information from the region of the artifact only.

8.5. NONDESTRUCTIVE TESTING

The field of nondestructive testing is a large one, and among its diverse techniques are many that use ultrasonic energy. These can conveniently be divided into two categories: echo location and imaging. The category we shall discuss is, of course, image formation. Traditionally, there have been two main methods for the generation of acoustic images, both of which rely on the principles of geometric optics. One is the use of a plane wave to cast a shadow of the object onto the detecting surface; the other is to use acoustic lenses to form an image.

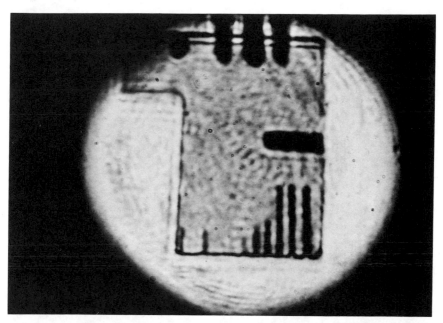

Fig. 8.20. As illustrated in this picture, optically opaque objects, such as this piece of plastic, are readily examined nondestructively to determine internal structure. In this case the eight holes drilled in from the bottom edge range in size from 2.24 mm to 0.35 mm in diameter. They are clearly imaged at an ultrasonic frequency of 5 MHz. The piece of plastic measures 7.5 cm from top to bottom.

 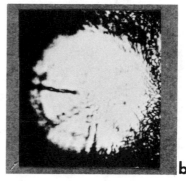

Fig. 8.21. These pictures demonstrate the focusing capability of the liquid-surface holography system and the imaging quality that is achieved in thick specimens of steel: (a) a 1.6 mm-diameter radial hole at a distance of 10 cm from the observer, through a 25 cm-long, 12.5 cm-diameter steel cylinder viewed axially, appears in focus at 4 o'clock. Defocused shadows are of holes at other depths. (b) a 3.2 mm-diameter radial hole at a distance of 17.8 cm from the observer, through the same 25 cm-long steel cylinder viewed axially, appears in focus at 9 o'clock in this ultrasonic image. Note that the hole at the 10 cm depth has now been defocused.

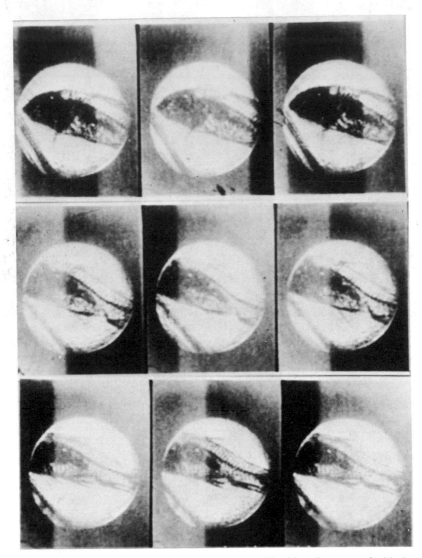

Fig. 8.22. The real time imaging characteristics of liquid-surface acoustical holography are illustrated in this figure which is a sequence of pictures taken from a 16-mm motion picture on which the acoustical images of a live goldfish were recorded.

8.5.1. Liquid-Surface Holography

Since in many cases we are concerned with relatively small trans-illuminated objects, liquid-surface holography can be used. The real-time feature of this system is particularly advantageous, since the object can be manipulated to obtain the best image. Often an internal flaw will appear in one orientation of the sample but not in another. It is also useful in the imaging of live biological samples. A selection of examples are shown in Figs. 8.20–8.22.

8.5.2. Scanning Holography

For applications where large structures are to be tested, or where trans-illumination is not possible, it is advantageous to use the scanning systems. Often, as in the case of a long structure, or where many identical pieces come off an assembly line, it is feasible to have the object do its own scanning by moving it on a conveyer belt through the acoustic beam. The source and receivers can therefore be stationary. This is exactly equivalent to one system described in Section 8.1.2 and redrawn in Fig. 8.23. Note that this particular application might be performed equally well with a liquid-surface system.

When the piece to be inspected is accessible from one side only, it is almost mandatory that a scanning system be used. The reason for this, in addition to low signal levels, is that time gating must be used to discriminate against the large echoes from the surfaces.

A problem peculiar to this application is the large change of velocity of sound from water to metal. Thus, when a flaw is imaged we must be careful to take into account the length of the water path if we wish to measure the position of the flaw within the piece. In addition, we must use the sound wavelength in the metal when calculating the size of the flaw.

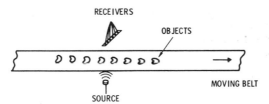

Fig. 8.23. Conceptual design for an on-line quality control inspection station for assembly-line flaw detection. This particular system utilizes the scanning object approach together with a line array of receivers.

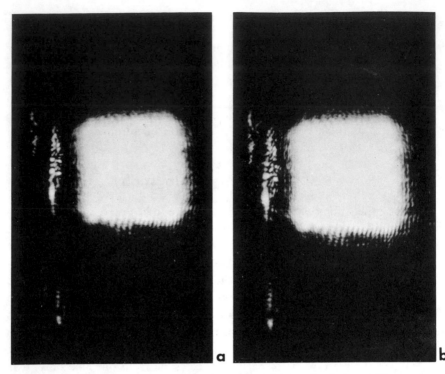

Fig. 8.24. Results from an industrial test performed for the Westinghouse Corporation. The object was to find and map a flaw in the section of a large machine rotor. The two photographs were taken at increasing depth within the image. The hologram was made with the scanning system, time-gated to reject reflections from the front and back surface of the specimen. The major flaw appears just to the left of the central order. The flaw was measured to be 10 cm long and about 12 mm wide at a mean depth of 6.8 cm in the casting. It was inclined at an angle with respect to the hologram aperture. Hence, when the image is explored in depth the best focus appears to shift from one end of the flaw to the other. (a) is a photo taken at a depth of 7.1 cm and (b) at 6.5 cm.

The large velocity difference between water and metal also means that considerable refraction occurs at the interface. This necessitates careful placement of the object in relation to the source and receiver to assure that sonic energy strikes the flaw and returns to the receiver. This problem can be overcome by using simultaneous source–receiver scan where the source is focused near the surface of the object. This assures that the sound enters the object at nearly normal incidence at all times. An example of a test performed for the Westinghouse Corporation on a section of a steel rotor is shown in Fig. 8.24. Note that as we scan through the image in the depth dimension, different parts of the flaw appear in focus. This indicates that the flaw lies on a slant with respect to the aperture.

8.6. INTERFEROMETRY

One of the major applications for optical holograpy has been a generalized interferometry. This is possible because the image formed by the hologram is perfect enough to be used as a reference against which to compare the distorted object. Therefore, if we make a hologram of the test object, replace it in the exact position in which it was made and view the stressed object through the hologram, we see fringes indicative of the displacements suffered by the surface of the object [14–16]. Alternatively, a double-exposure hologram may be made with the object being stressed between exposures [17,18]. For transient stresses, double-pulsed lasers may be used [19].

A different type of interferometry may be performed by changing the frequency of the source between exposures. This provides an image having contours of constant depth superimposed on it, allowing accurate depth measurements to be made [20]. The contour interval is shown to be $\lambda^2/2\Delta\lambda$

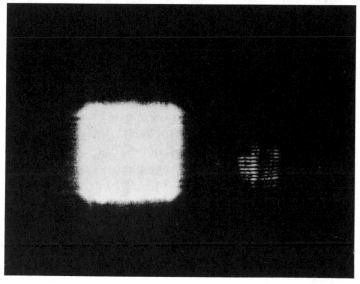

Fig. 8.25. This is a photograph of the image of a hologram of an aluminum sheet which was rotated by 1° between two scans. This illustrates the possibility of perfoming interferometry at longer wavelengths so that measurements of engineering accuracies can be easily performed.

where λ is the mean wavelength and $\Delta\lambda$ the wavelength change between exposures.

It is well known that interferometry is an extremely accurate measurement technique, with fringes occurring for displacements less than a wavelength. For many engineering measurements this is too sensitive since distortions of general interest produce fringe densities much too large to record. Consequently, many clever schemes have been devised to desensitise the technique [21]. Most of these schemes are related to the Moiré fringe method [22]. One method that has generally been overlooked is that of simply using longer wavelengths.

It is entirely feasible to repeat all the various interferometric arrangements used at optical wavelengths with acoustical waves. Since the wavelengths in water are 250 to 500 times longer than light wavelengths, displacements of engineering interest can be measured. Figure 8.25 is an example of the reconstructed image from a double-exposure acoustical hologram made at 10 MHz. Note the high-contrast fringes. Similarly, experiments could be performed using two acoustical wavelengths to produce a contoured image. For low-contrast images, such as might be formed of the ocean bottom, contours would be invaluable for interpretation of ocean bottom topography.

8.7. SUMMARY

In this chapter we have described a number of proposed applications for acoustical holography, some of which we have demonstrated in the laboratory. Actually bringing these ideas to practical use will require considerable engineering effort, particularly the large-scale application to undersea viewing and to seismic exploration. Nevertheless, the latter is already being attempted by a large oil exploration firm [23].

Although we have for the most part restricted our discussion to liquid-surface and scanned holography, any of the other detection and recording methods, when more fully developed, can also be used. In general, the scanning systems stand apart. The other methods, like photographic film, Bragg diffraction, ultrasound camera, particle cell, liquid crystals, and temporal reference all fall into the liquid-surface category in terms of aperture size. These methods, until they prove themselves superior to the liquid-surface system, will only remain interesting research topics.

REFERENCES

1. D. C. Greene and B. P. Hildebrand, Acoustic holograms for underwater viewing, *Ocean Industry* **4**(9):43; **4**(10):47 (1969).
2. C. E. Cook and M. Bernfeld, *Radar Signals*. Academic Press, New York (1967).
3. V. C. Anderson, Digital array phasing, *J. Acoust. Soc. Am.* **32**:867 (1960).
4. D. Silverman, Wavelet reconstruction process for sonic, seismic, and radar exploration, U.S. Patent No. 3,400,363, 1968.
5. J. B. Farr, *Earth Holography, A New Seismic Method*, Paper presented at 38th annual meeting of the Society of Exploration Geophysicists, Denver, Colorado, 1968.
6. J. M. Crawford, Wm. E. N. Doty, and M. R. Lee, Continuous signal seismograph, *Geophysics* **25**:95 (1960).
7. D. Gabor, G. W. Stroke, R. Restrick, A. Funkhouser, and D. Brumm, Optical image synthesis (complex amplitude addition and subtraction) by holographic Fourier transformation, *Phys. Lett.* **18**(2):116 (1965).
8. B. P. Hildebrand, *Holographic Subtraction*, Report no. ODD-68-8, Battelle-Northwest, Richland, Washington, 1968.
9. J. Upatnieks, A. VanderLugt, and E. N. Leith, Corrections of lens aberrations by means of holograms, *App. Opt.* **5**:589 (1966).
10. B. P. Hildebrand, *Acoustical Holography for Nuclear Instrumentation*, Report No. BNW-SA-2017, Battelle-Northwest, Richland, Washington, 1968.
11. K. T. Dussik, Über die Möglichkeit hochfrequente mechanische Schwingungen als diagnostisches Hilfsmittel zu verwanden, *Z. ges Neurol. Psych.* **174**:153 (1942).
12. D. N. White, *Ultrasonic Encephalography*, Hanson and Edgar Inc., Kingston, Ontario (1970).
13. L. Weiss and E. D. Holyoke, Detection of tumors in soft tissues by ultrasonic holography, *Surg. Gyn. Obstet.* **128**(5):953, 1969.
14. B. P. Hildebrand and K. A. Haines, Interferometric measurements using the wavefront reconstruction technique, *Appl. Opt.* **5**:172 (1966).
15. R. J. Collier, E. T. Doherty, and K. S. Pennington, Applications of moire techniques to holography, *App. Phys. Lett.* **7**:223 (1965).
16. J. M. Burch, 1965 Viscount Nuffield Memorial Paper, *The Production Engineer* **44**: 431 (1965).
17. K. A. Stetson and R. L. Powell, Hologram interferometry, *J. Opt. Soc. Am.* **55**:1570 (1965).
18. R. E. Brooks, L. O. Heflinger, and R. F. Wuerker, Interferometry with a holographically reconstructed comparison beam, *Appl. Phys. Lett.* **7**:248 (1965).
19. R. F. Wuerker and L. O. Heflinger, Pulsed laser holography, in *The Engineering Uses of Holography*, E. R. Robertson and J. M. Harvey (eds.), Cambridge University Press (1969).
20. B. P. Hildebrand, The role of coherence theory in holography with application to measurement, in *The Engineering Uses of Holography*, E. R. Robertson and J. M. Harvey (eds.), Cambridge University Press (1969).
21. J. Der Hovanesian and J. Varner, Methods for determining the bending moments in normally loaded thin plates by hologram interferometry, in *The Engineering Uses of Holography*, E. R. Robertson and J. M. Harvey (eds.), Cambridge University Press (1969).
22. P. S. Theocaris, *Moire Fringes in Strain Analysis*, Pergamon Press, New York (1969).
23. Pan American Petroleum Company, 1969 Annual Report.

Index